Environmental Ec

This book helps those concerned with environmental issues to firmly grasp relevant analytical methods and to comprehend the thought process behind environmental economics. It does so by drawing from specific environmental issues and at the same time providing commentary that facilitates understanding. This text is replete with detailed explanations that are fundamental to a thorough understanding of the components and significance of environmental economics.

Environmental Economics aims to clarify the mechanisms that give rise to environmental problems by approaching environmental issues from an economic perspective. At the same time, its purpose is to identify specific countermeasures that could resolve existing environmental challenges. To this regard, chapter content is arranged in a manner that renders pertinent topics and issue areas approachable and comprehensible to a general audience.

Shunsuke Managi is the Distinguished Professor of Technology and Policy in the Urban Institute & School of Engineering at Kyushu University, Japan, while also holding positions as an Adjunct Professor at QUT Business School, and the University of Tokyo. He is an editor of *Environmental Economic and Policy Studies*, a lead author for the Intergovernmental Panel on Climate Change, is the author of *Technology, Natural Resources and Economic Growth: Improving the Environment for a Greener Future* and editor of *The Economics of Green Growth*.

Koichi Kuriyama is the Professor of Agricultural and Resource Economics at the Kyoto University, Japan. His research has focused on the evaluation of the environment including forest and ecosystem services. He is the author of 'A latent segmentation approach to a Kuhn-Tucker model: An application to recreation demand' (with W.M. Hanemann and J.R. Hilger), published in the *Journal of Environmental Economics and Management*.

Routledge Textbooks in Environmental and Agricultural Economics

For a full list of titles in this series, please visit https://www.routledge.com/series/ TEAE

Environmental Economics

Shunsuke Managi and
Koichi Kuriyama

LONDON AND NEW YORK

First published 2017
by Routledge
2 Park Square, Milton Park, Abingdon, Oxon OX14 4RN

and by Routledge
711 Third Avenue, New York, NY 10017

Routledge is an imprint of the Taylor & Francis Group, an informa business

British Library Cataloguing in Publication Data
A catalogue record for this book is available from the British Library

Library of Congress Cataloging in Publication Data
Names: Managi, Shunsuke, author. | Kuriyama, Koichi, 1967- author.
Title: Environmental economics / by Shunsuke Managi and Koichi
    Kuriyama.
Description: Abingdon, Oxon ; New York, NY : Routledge, 2017. | Series:
    Routledge textbooks in environmental and agricultural economics ; 17 |
    Includes bibliographical references and index.
Identifiers: LCCN 2016013850| ISBN 9781138960688 (hardback) |
    ISBN 9781138960695 (pbk.) | ISBN 9781315467337 (ebook)
Subjects: LCSH: Environmental economics. | Economic development—
    Environmental aspects.
Classification: LCC HC79.E5 M34547 2017 | DDC 333.7—dc23
LC record available at https://lccn.loc.gov/2016013850

ISBN: 978-1-138-96068-8 (hbk)
ISBN: 978-1-138-96069-5 (pbk)
ISBN: 978-1-315-46733-7 (ebk)

Typeset in Times New Roman
by Swales & Willis Ltd, Exeter, Devon, UK

Visit the companion website: www.routledge.com/cw/shunsuke

Contents

Figures

Tables

Preface

Purpose of this book

This book helps those concerned with environmental issues to firmly grasp relevant analytical methods and to comprehend the thought process behind environmental economics. It does so by drawing from specific environmental issues and at the same time providing commentary that facilitates understanding. This text is replete with detailed explanations that are fundamental to a thorough understanding of the components and significance of environmental economics.

Environmental Economics aims to clarify the mechanisms that give rise to environmental problems by approaching environmental issues from an economic perspective. At the same time, its purpose is to identify specific countermeasures that could resolve existing environmental challenges. Nowadays, we come into direct contact with the environment in a number of facets. And, global environmental issues such as climate change, pollution problems such as air and water contamination, ever-worsening waste issues inherent in mass consumption societies, ecosystem destruction brought about by dam development and other public works projects, environmental impacts arising from international trade and the globalization of economies, and various other environmental challenges are ever-present. This book aims to draw from these specific environmental issue areas while providing an easy-to-understand commentary on the analytical methods and ways of thinking inherent in environmental economics.

Special features

This book contains the following special features which set it apart from conventional texts on environmental economics.

1. Thought has been put into providing information that can be read and understood without specialized economics knowledge.
2. Explanations are provided while drawing from concrete environmental issues such as climate change and waste problems.
3. Up-to-date research content is introduced.
4. Figures and commentary illustrate without overly complex formulas
5. Related topics are introduced in "Learning point" sections.

The first unique feature of this textbook is that due consideration has been put into providing information that can be read and understood without specialized economics knowledge. People interested in environmental economics, including those without formal education in economic studies, who have learned about environmental issues through other disciplines, those who have taken on environmental challenges in administrative and government positions, and other full-time workers who have had to confront environmental challenges, can all make use of the content within.

Furthermore, in order for those unfamiliar with economic studies to be able to read and grasp this text smoothly, technical terms have been omitted to the greatest extent possible, and special care has been put towards providing detailed explanations so that even beginners can quickly comprehend the presented materials.

Secondly, this textbook is novel in that it draws from climate change, waste challenges, and other specific examples as much as possible in order to explain the ideals that underlie environmental economics. Generally speaking, economics textbooks focus largely on explaining the fundamentals of economics, but this book focuses not only on said fundamentals of environmental economics, it also introduces the particulars of coping with existing environmental problems. Even those who have difficulty comprehending the abstract economics perspectives provided in contemporary texts should be able to concretely understand economic thought processes by reading this book and comparing the detailed depictions of actual environmental problems.

It is also worth noting that this book introduces up–to-date research content. For example, environmental value assessments, relationships between businesses and the environment, and other new research themes have been heavily researched and become core components of environmental economics in recent years. Through the present, texts on environmental economics seldom bring up the specific details of these research themes. However, this work provides comprehensible explanations of the latest research results. Even for those who have already begun studying environmental economics with other texts, there should be plenty to learn from the new research content included in this book.

The fourth original aspect of this work is that it employs figures and charts for clear explanations without over-relying on difficult formulas. Even for those who have ample experience with the numerical formulas that are commonplace in economic studies, and especially for students who are only just beginning to learn economics or others who are not well versed with economics, it takes a fair amount of time to be able to quickly and efficiently grasp deeper meanings through formulas. Thus, in most cases, this book depends on its text and figures in place of formulas when providing explanations. And where figures are used, they are accompanied by thorough descriptions of how to read them and other particular information.

Finally, "Learning point" corners have been appended to introduce pertinent topics for understanding the thinking behind environmental economics. For example, by using simplified economic experiments as means to explain economic thinking, and by introducing how real-life environmental procedures are applied via environmental economics, this book sheds light on realistic, relevant themes.

How this book should be read

The information provided in this book corresponds with a full year of lecture material. The first half, consisting of Chapters 1 through 4, would generally be covered in the first semester, while Chapters 5 through 7 would be covered in the second semester. As Chapters 4 through 7 include ever-evolving content, in some cases, abbreviating or omitting material may be called for. In half-year courses, addressing either the first or second half of the book, exclusively, may be appropriate.

The beginning of each chapter provides a summary of what one could expect to learn from content contained within subsequent sections. One section includes content appropriate for a single lecture, each contributing detailed descriptions of distinct topics. It does not matter whether the sections are read in order or whether only those related to topics of interest are the study focus. Important concepts are provided as keywords at the beginning of each section. At the end of each chapter, one should use the summaries provided to review the section content.

Furthermore, review problems are also provided at the end of each section. These are ideal for readdressing and confirming one's understanding of the content addressed in the section.

Moreover, this book considers the needs of a wide range of readers. Those interested in environmental economics include not only economic studies majors, but also consist of people from a diverse array of academic majors that cover environmental issues, policy makers and administrative workers, business people who are involved in environmental policies, and even a host of people from the general populace. This book is intended to inspire an enjoyable study experience for a far-reaching and diverse set of learners, all the while equipping them with the fundamentals necessary to apply environmental economics.

The writing included in this book is the result of broad-based collaboration with a great number of individuals. While the authors do themselves teach environmental economics in university lectures, close attention was paid to student opinions regarding lecture content when drafting this text. Furthermore, the successful research results acquired from cooperative experiments with numerous other researchers are included in the examples provided throughout the book. Finally, we wish to thank Clarence Tolliver III for his great help in drafting the text.

Shunsuke Managi
Koichi Kuriyama
January, 2016

Introduction

From climate change and other environmental problems that take place on a global scale, to waste disposal and other relatively close-to-home issues, nowadays, people across the globe must contend with a variety of challenges related to the environment. How do these environmental issues arise in the first place? Furthermore, what is needed to solve environmental problems? Environmental economics approaches the previously mentioned environmental quandaries from economic perspectives.

Consider, for example, the link between the environment and economics in the context of climate change. Climate change is a phenomenon through which the Earth's surface temperatures rise hand in hand with greater economic activity. That is, as vast amounts of oil and coal fuel ever-expanding economic processes, large loads of carbon dioxide (CO_2) and other greenhouse gases are emitted into the atmosphere, resulting in rising global surface temperatures. As climate change intensifies, rising sea levels, floods, droughts, and other disasters become more prevalent. Moreover, changes among the Earth's natural conditions are expected to affect agricultural productivity and possibly lead to the wildlife extinctions and ecosystem collapses across the planet. Due to the scale and severity of these risks, coping with climate change has become an unavoidable, pressing issue.

Unfortunately, applying policies and other methods to tackle the climate change issue has been an incredibly challenging task. The primary reason for this is that there are costs to preventing climate change. Various strategies, from employing wind power and other initially expensive electricity generation technologies, to implementing reforestation programs for carbon dioxide sequestration, are essential to reducing greenhouse gas emissions. The downside of such initiatives, however, is that they tend to render enormous costs.

From this perspective, it is clear that the environment cannot be preserved for free. For, first and foremost, there are no direct benefits to businesses even if they do act to safeguard it. Moreover, since the natural environment has no market value (i.e. the environment is handled as if it is "free of charge"), the economic constructs that lay the groundwork for functioning society fail to properly assess environmental worth, leading to precipitous, widespread environmental degradation. What this means, then, is that many environmental problems originate from or are spurred on by the economic systems at the core of society (see Figure I.1).

Environmental challenges must therefore be assessed through an economic lens.

Mechanisms of the economy

When economic activity is free to yield pollution:

(1) Businesses earn profits
(2) Environmental degradation continues over time.

When economic activity is adjusted to protect the environment:

(1) Businesses do not earn profits
(2) Environmental preservation costs emerge
(3) Environmental preservation aims are eventually abandoned.

FIGURE I.1 Environmental issues and the economy

What is environmental economics?

Environmental economics is the study of environmental issues through economic assessments. Figure I.2 highlights the salient issue areas of environmental economics, which are mainly comprised of the following three topics:

1. "Why do environmental issues arise?",
2. "What is needed to resolve environmental issues?", and
3. "How can a society capable of both environmental conservation and economic development be realized?"

The first of these fundamental topics relates to the economic mechanisms that give rise to environmental issues and sheds light on the origins of said issues. As previously mentioned, economic systems that treat the environment as if it were cost-free lie at the root of environmental issues. More specifically, phenomena known by economists as *market failures*. Through environmental economics, various real-life environmental predicaments, from climate change to waste disposal issues, are targeted to examine the forms that market failures take.

The second principle concern of environmental economics is to determine specific policy measures for dealing with environmental issues. Traditionally, local environmental policies are comprised of various *environmental regulations*, compelling businesses to moderate their pollution emissions. While they have been lauded for their efficacy in coping with air and water contamination, conventional measures of this sort are rather ineffective when it comes to dealing with climate change, waste disposal, and other more recent challenges.

To this end, greater attention is given to environmental taxes, emissions trading systems, and similar economic measures. Take *environmental taxes*, which impose fees on pollution, as prime examples of effective economic measures. Beyond merely

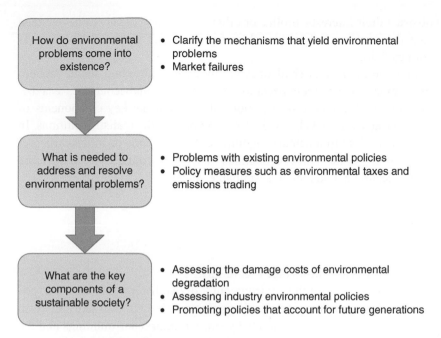

FIGURE I.2 A breakdown of environmental economics subject matter

levying tax fees, environmental taxes incentivize businesses to reduce their emissions since doing so translates into reduced costs through fewer taxes imposed by the system. Similarly, emissions trading and other economic measures also consider the economic behavior of businesses and consumers in order to make pollution reductions of personal interest to various actors. These, as well as other policy processes, are important subjects in the realm of environmental economics.

The third and final central topic involves indicating what is needed to make the transition to sustainable society. Sustainable society takes the needs of people in the present generation, as well as those of our children's, grandchildren's, and other future generations, into account. It is a society that aims for harmony between environmental conservation and economic growth.

In order to realize sustainable society, it is first necessary to evaluate the environment and the economy with the same criteria. For instance, many predict that the the floods and droughts brought about by climate change will be severe and costly. Furthermore, widespread wildlife extinction and other ecosystem impacts that climate change will have must not be overlooked. If costs of environmental destruction are not made clear, then concrete environmental policies cannot be advanced. Therefore, environmental economics provides methods to assess the monetary price of the environment to be used within modern environmental policy frameworks.

Furthermore, businesses ought to play significant roles in realizing a sustainable society. Now more than ever, a great number of businesses actively engage in environmental initiatives in the form of corporate social responsibility. However, businesses are less likely to fully adopt environmental initiatives that fail to tie into or serve to obstruct their interests, profits, or other gains. Such connections between businesses and the environment are thereby of prime importance in the environmental economics field.

Ever-worsening environmental challenges are met with increasing social concern for realistic resolutions and greater expectations of environmental economics as a discipline. In fact, the field is now tasked to not only explain the key components of environmental issue analysis, but also with providing pragmatic, realistic solutions. In this way, the mandate of environmental economics extends beyond its initial calling of questioning the viability of the market economics-based industrial activity to the more inclusive goal of creating a road map to sustainable society

Book layout and contents

The layout of this book is as follows (see Figure I.3):

"Chapter 1: Our lifestyles and the environment," draws from waste disposal, climate change, and other environmental problems that are tied into our daily lives, in order to highlight the inherent connection between our natural surroundings and the economy. The downside to economic growth is that it usually yields great quantities of waste, resulting in severe waste-related issues. All the while, widespread, high-volume energy consumption greatly contributes to ever-increasing CO_2 gas emissions and exacerbates the climate change situation.

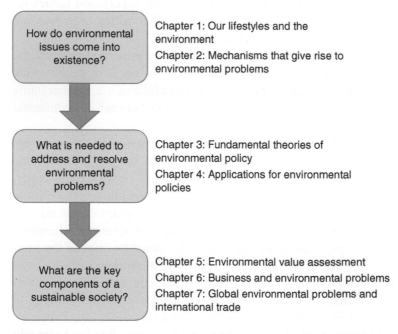

FIGURE I.3 This book's approach to environmental economics subject matter

"Chapter 2: Mechanisms that give rise to environmental problems," provides a closer look at the origins of the previously mentioned environmental issues through an economic lens. Market mechanisms are incapable of efficiently functioning for a natural environment that possesses no defined market value (e.g. an environment that is "free"). Further still, there are costs involved in protecting the environment, and people may choose not to bear the burden of these costs if someone else might in their stead. What this means is that market economics, when relied upon solely, could not only fail to produce adequate solutions, but could also exacerbate existing problems.

"Chapter 3: Fundamental theories of environmental policy," provides an overview of the fundamental economic theory behind environmental policies. Environmental policies include both direct regulations and more indirect, economics-centered control measures (environmental taxes, emissions trading systems, etc.). A detailed account of these policies is included in this chapter.

"Chapter 4: Applications for environmental policies," depicts how environmental policies put environmental economics methodologies to practical use. Waste policy and climate change policy precedents are cited to show how direct regulations and economic control measures are employed.

"Chapter 5: Environmental value assessment" demonstrates how to assess the monetary unit value of the environment. The first method for achieving this is to make indirect assessments by observing the economic behavior of individuals. The second method is to make direct assessments by asking individuals about their value perceptions.

"Chapter 6: Businesses and environmental problems," explains the role of businesses in environmental policies. Due to growing public concern for environmentally friendly business practices, businesses must now take it upon themselves to demonstrate their environmental awareness and impact to consumers and investors. And, in addition to crafting their own unique company strategy to achieve the above, businesses must also firmly grasp the costs and effects of the environmental countermeasures they employ.

"Chapter 7: Global environmental problems and international trade," offers a focused analysis of the links between international trade and environmental issues, particularly in the context of rapidly expanding, globalizing economies. When analyzing the nature of certain environmental issues, it is extremely important to note how they respectively relate to present and future generations. This chapter expands on these themes and how the concept of sustainable development is related to solving global environmental problems.

LEARNING POINTS

1. Are economic development and environmental conservation compatible with each other?
2. An economic experiment with waste problems
3. Preventing climate change

Our lifestyles and the environment

Chapter overview

This chapter depicts how our lifestyles are interconnected with the environment. From waste disposal and other matters directly linked to our personal lives, to such phenomena as climate change that exert influence to all people across the planet, our lifestyles are closely connected to many environmental issues. Citing these sorts of precedent-setting examples, this chapter will explain the economic systems that underlie said environmental problems.

The first aim is to reflect on the relationship between economic growth and environmental issues. The greater the economic growth, the more natural resources are used, and subsequently, the more severe environmental pollution becomes. Such trends also point to the danger of surpassing the planetary boundaries that are quintessential to life (through such phenomena as ocean acidification, ozone depletion, etc.). Before arriving at these conditions (which would also place checks on further development), it is important to consider what is needed to make the transition to a sustainable society.

The second aim of this chapter is to consider various initiatives for coping with waste disposal issues. Since there is limited landfill space, were waste output to continually rise, disposal space would eventually be depleted. To that regard, promoting waste reduction and recycling schemes is imperative to achieving sustainable society.

Finally, the third aim of this chapter is to closely assess the climate change predicament. Although climate change is largely caused by fossil fuel and energy consumption in the present, the majority of its dire effects will only be realized some hundreds of years in the future. To avoid these long-term consequences, members throughout the international community must cooperate in adopting efficient policies.

This chapter also includes a simplified economic experiment, providing a first-hand glimpse at the severity of waste issues.

Chapter content

Section 1.1—With economic development, greater amounts of natural resources are used. Additionally, as pollution becomes more severe, humankind runs the risk of surpassing the planetary boundaries that are essential for life. It is possible to achieve both economic growth and environmental conservation if government action or market mechanisms can generate effective environmental regulations. However, it is important to consider the long-term ramifications of economic growth in the present, as many of the harmful side effects of current economic activities (e.g. climate change) will not surface until many years into the future.

Section 1.2—As mass consumption societies generate profuse quantities of waste, diminishing landfill space is an increasingly critical issue. To date, governments around the world have enacted various recycling and waste related laws to address these concerns. With that said, recycling comes with considerable costs, and if recycled goods fail to sell, then the company overseeing the recycling operations cannot accrue profits. Cutting waste disposal costs to the greatest extent possible, a critical step in the transition to sustainable society, is achievable through adherence to the tenets of economic efficiency.

Section 1.3—Greater fossil fuel energy consumption, spurred on by consistent economic growth, leads to greater climate-change-related risks. Fossil fuel energy consumption is based on the balance between fossil fuel supply and demand. Yet these market mechanisms fail at coping with climate change since they traditionally do not take the effects of climate change into account. To overcome this, developed countries have traditionally joined together to design emissions reduction goals and adopt multilateral protocols for tackling climate change. Even still, global policies that incorporate developing countries remain at the core of climate change prevention policy discourse.

Section 1.1: Economic development and environmental issues

Economic development and mass consumption society

The global economy is expanding rapidly (see Figure 1.1.1). In the 50 years between 1950 and 2000, it scaled up by a multiple of 7. Recent years are marked by considerable growth in Asian markets in particular. In the most basic sense, economic development leads to increased prosperity, and people living in countries that have succeeded at bolstering their economies to certain levels can expect high or even affluent living standards. Thus, many developing countries seeking this sort of prosperity hold economic growth to the highest level of priority.

Millions of USD

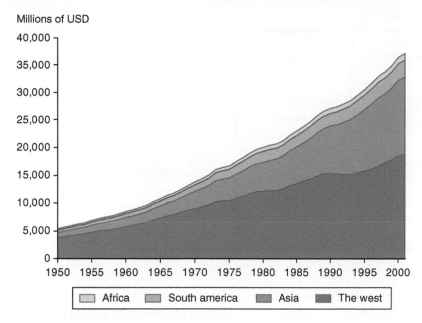

FIGURE 1.1.1 Economic development trends

Source: *Development Centre Studies of The World Economy Volume 2: Historical Statistics*, OECD, 2006

However, rapid economic growth gives rise to mass consumption societies, characterized by high production, consumption, and waste levels. Without a doubt, mass production has made it possible to supply industrial goods at cheaper prices than ever before. This has in turn contributed to higher standards of living throughout society. Take the proliferation of the automobile as a prime example of this trend. In early 1900s America, the automobile was a luxury commodity, affordable only by more affluent classes who made up a mere fraction of society. But upon the introduction of the mass production technologies that were integral to the Ford Motor Company assembly line, production costs fell sharply, rendering automobiles affordable to the general public. From then on, automobiles quickly became commonplace, intrinsic to the everyday lives of the American people.

Notwithstanding, widespread diffusion of the automobile has also led to increased traffic congestion and a number of serious health risks. Moreover, automobile carbon dioxide (CO_2) emissions act as climate change catalysts, and as such, increasing automobile numbers coincide with the elevating climate change risks. From this example, mass consumption societies have their benefits as well as their drawbacks. For, while they do yield the boons of economic and industrial output, they also have a considerable negative impact on the environment.

Figure 1.1.2 illustrates the connections between the environment and the economy. The economy consists of the production and consumption of goods and services. In order to achieve this, oil, minerals, and other resources must first be

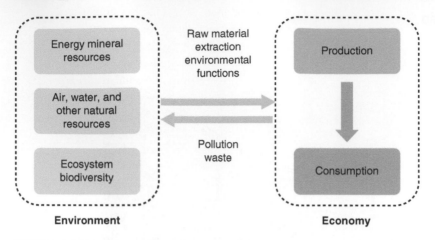

FIGURE 1.1.2 Links between the environment and the economy

extracted from the Earth. Yet crude oil and many important mineral resources are finite and cannot easily be replenished. Thus, humankind would be confronted with quite a precarious economic situation should the buried reserves of these materials be used to exhaustion.

Beyond the consideration of resource reserves, the air and water pollution emitted by factories when goods are produced are also aspects of the economy that should not be overlooked. When production levels are low, pollution emissions are also low, so naturally occurring environmental cleansing cycles remain capable of processing pollutants and preserving the purity of the natural environment. Conversely, when production levels surge hand in hand with economic development, factory pollution emissions also rise. In such cases, air and water contamination levels outstrip the capacity of natural purification cycles to deal with them, leading to adverse pollution effects.

At the same time, consumers and the environment are intimately connected with one another. The high-volume consumption that is inherent in mass consumption societies yields a great deal of trash and other waste. Limited landfill space fuels debates concerning where waste ought to be buried and processed. Japan, as a country with minimal national territory, lacks the land space necessary for waste disposal processing. Thus, its current mass consumption, mass waste production trends have exacerbated an already serious waste disposal conundrum.

From what was said above, mass consumption societies, though forged in the fire of high economic attainment, can be viewed as the harbingers of resource depletion, severe air and water contamination, and grand-scale waste disposal challenges.

The Limits to Growth

If economic growth trends continue as they are, and countries throughout the world realize economic standards comparable to developed nations, the Earth's planetary boundaries will likely be exceeded. Outlines known as *ecological footprints* are used

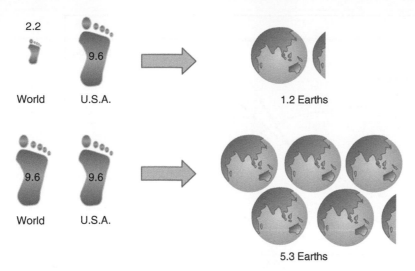

FIGURE 1.1.3 Ecological footprint

Source: *Living Planet Report*, WWF, 2006

to indicate the effects that economic activity has on the ecosystem. According to the *Living Earth Report: 2006 Edition* published by the World Wildlife Foundation, the land space that would be necessary to maintain current economic standards of the entire globe amount to 2.2 hectares per capita. Comparatively, people in the U.S. alone would require 9.6 hectares per capita. Thus, if the entire world aimed to develop to U.S. standards, four Earths would be necessary (see Figure 1.1.3). This clearly surpasses the tolerable limits of the Earth, rendering it impossible.

Supposing, however, economic development were to continue along current trends and planetary boundaries were exceeded, what type of conditions would this present? A report titled *The Limits to Growth* published by the Club of Rome in 1972 warned that economic growth would reach a limit as resources deplete and the environment degrades due to industrial activities. Figure 1.1.4 indicates a number of such growth limits. Since economic development leads to extensive use of crude oil and mineral resources, their stocks would exhaust eventually. As economic processes are largely dependent upon these finite, non-renewable resources, as they deplete, the speed of development would slow, peak, and then rapidly decline. Beyond all of this, health pandemics that originate from pollution, shortages in food supply, and other hardships brought about by highly polluting economic activities, would result in a downturn in the human population. Thus, according to this view, a bleak future awaits the world should economic development continue at its current pace.

Economic growth and environmental preservation

Of course, there are many criticisms of this extreme scenario. The first is with regards to resource depletion issues. Since productivity is increasing alongside advances

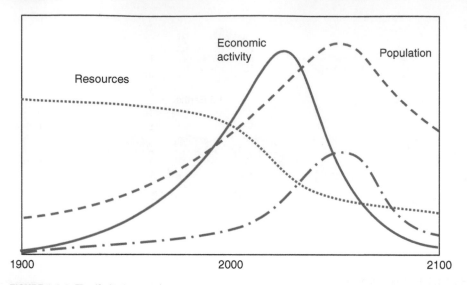

FIGURE 1.1.4 The limits to growth

Source: Meadows et. al, *Beyond the Limits*, Diamond, 1992

in technology, resource reserves will not necessarily be rapidly and completely exhausted. For instance, during the oil shocks of the 1970s, it was believed that oil reserves would last a mere 30 years or so thereafter. However, with the evolution of oil extraction technologies, it has become possible to extract oil from regions that were previously thought of as economically unviable. Modern times are thus not characterized by depleted oil reserves as was once predicted.

Secondly, *The Limits to Growth* does not take the price adjustment features of market mechanisms into consideration. *The Limits to Growth* assumed that resources would be continually used through exhaustion. Yet in reality, as resource reserves begin to dwindle, market mechanisms cause their prices to soar. For example, even if production levels were curtailed in the event of perceived oil reserve depletion, this shortage of supply would cause oil prices to rise. This would in turn mean that oil could no longer be procured and used as profusely as before, impelling consumers to purchase vehicles with better gas mileage and live more energy-efficient lifestyles. Another due consideration is that wind energy and other natural energy resources could potentially supplant conventional thermal energy should their related technologies be properly invested into and diffused. Therefore, the existence of price adjustment features within market economies makes it difficult to imagine continuous, mass-quantity exploitation of natural resources to the point of depletion and economic collapse.

A third criticism of the above concerns environmental pollution, as damages could be curbed through emissions regulations imposed by the government. For example, during the post-WWII period of rapid economic growth in Japan, there were various outbreaks of pollution-induced dangers, including Minamata Disease. Once they caught the public eye, the government implemented factory emission regulations, leading to great reductions in air and water pollution damages. Thus, through state environmental regulations, society can steer clear of severe pollution and its devastating effects.

The fourth criticism is that as economies develop, higher income standards tends to foster greater public concern for environmental issues, resulting in policies enacted to address said concerns. While developing countries tend to value development over environmental conservation, developed countries reach a point where they are unable to turn a blind eye to environmental protection policies in the face of heightened public awareness and concern.

Thus, as suggested by all of the above criticisms of the *The Limits to Growth,* it is certainly possible to imagine a world in which economies continue to thrive as negative environmental impacts abate.

Sustainable development

When effective environmental regulations are put to work via market mechanisms or government action, the gloomy visions of the future described in *The Limits of Growth* begin to seem avoidable. That is not to say, however, that that market mechanisms and government ordinances will always work well. The climate change issue demonstrates this clearly. Despite international awareness, current CO_2 and other greenhouse gas emissions continue to escalate. One of the reasons for the dearth of progressive climate change policy is that there exists no market price for the environment, rendering market mechanisms futile. Policy measures that promote wind power and other natural energy resources, as well as substantial financial contributions, will be needed to avert climate change. What's more, even in cases where businesses adhere to climate change policies, since there is no universally defined monetary value of the atmosphere and many other aspects of nature, businesses have little or no incentive to act proactively for their preservation. This inevitably serves as an impediment to avant-garde environmental policy adoption among firms.

Moreover, while government measures are essential to confronting climate change in the present, the climate change-related damages often do not arise in the present. In other words, a unique characteristic of climate change is that many of its negative effects will only come about some hundreds of years into the future. Governments that fail to include this sort of long-term vision with regards to the effects of climate change will inevitably develop ineffective policies. In these circumstances, where neither market mechanisms nor government regulations function properly, it would become impossible to avoid the risks of climate change. Society would fail to approach the critical turning point on the environmental Kuznets curve (refer to the **Learning point**), and greenhouse gas emissions would continually rise. In the same way depicted by *The Limits to Growth*, the economy would inevitably fall asunder in the wake of climate change.

Thus, when confronting such global environmental issues as climate change, it is imperative to consider both present and future generations. At some point along the course of economic advancement, humankind must adopt new, more forward thinking to processes inherent in *sustainable development* (refer to Chapter 7, Section 7.3 for details).

LEARNING POINT: ARE ECONOMIC DEVELOPMENT AND ENVIRONMENTAL CONSERVATION COMPATIBLE WITH EACH OTHER?

When income levels in a given society are low, individual and business entities in said society seek greater income through economic development. In turn, the natural environment suffers damages that coincide with these aims. However, stresses on the environment can gradually be alleviated when governments implement a successful environmental policy. This relationship between income levels and environmental pollution can be depicted by the reverse U-shaped curve known as an *environmental Kuznets curve* (refer to Chapter 7, Section 7.3 for further details). The environmental Kuznets curve gives meaning to the potential coexistence of economic development and environmental conservation, and it diverges completely from the portrayals found in *The Limits to Growth*.

However, it is important to note that depending on the type of environmental pollutant involved, the environmental Kuznets curves may or may not be empirically observed. Sulfur dioxide (SO_2), a substance known to cause atmospheric pollution, has been observed to form environmental Kuznets curves throughout multiple developed countries. On the other hand, carbon dioxide (CO_2), a major catalyst of climate change, has not portrayed this, regardless of the high-level economic attainment achieved by the emitting country. Thus, at present, it is difficult to imagine an environmental Kuznets curve for CO_2.

SUMMARY

Economic development leads to massive resource consumption and environmental pollution. As pollution grows more severe, so too does the risk of surpassing the planetary boundaries that sustain life. It is possible to realize both economic progress and environmental preservation simultaneously if market mechanisms and government regulations function effectively. However, as is evident from such phenomena as climate change that are created in the present yet pose greater risks to future generations, it is imperative to consider the effects that current economic activities will have in the long run.

REVIEW PROBLEMS

1. Look up Japan's domestic SO_2 emissions and per capita GDP data and create a scatterplot graph. Then, explain the relationship between economic growth and SO_2-induced environmental contamination.
2. Explain why CO_2 is not applicable to an environmental Kuznets curve.

3. In a tropical region, rainforest destruction is taking a toll. Assuming the economic standards of local residents were to improve through this activity, would this lead to the tropical rainforest's restoration or its continued destruction? Why?

Section 1.2: Waste problems and sustainable society

What are waste issues?

Waste disposal is an issue that is closely connected to our everyday lives. In mass consumption societies, after large quantities of goods are produced and consumed, copious amounts of waste are yielded. But because there is limited space for waste disposal facilities, at some point, room for such locations will run out. This is an especially weighty consideration for countries like Japan which possess limited territory and few places that are suitable for waste disposal.

Figure 1.2.1 depicts waste disposal trends. Note the rapid increase in waste emissions in Japan during the latter half of the 1960s, the same period in which it entered a phase of rapid economic growth. From that period on, emissions levels climbed marginally over periods of lower growth, and today, emission fluctuations have largely leveled out (though emissions are not actually decreasing).

Waste can be generally classified into two distinct forms: *general waste* that is produced by common households, and *industrial waste* that is generated in industrial processes. First, let's explore a few topical issues surrounding general waste. The burnable materials among ordinary household trash are incinerated, and their

Units: 1,000 tons

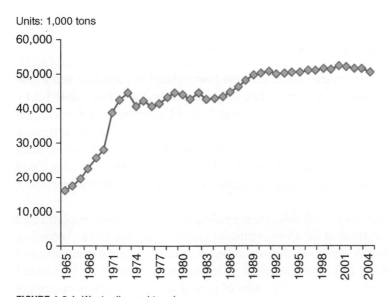

FIGURE 1.2.1 Waste disposal trends

Source: *Environmental Statistics*, Ministry of the Environment

ash remains are buried at *final disposal sites* (e.g. landfills). Glass and plastic bottles are recycled, and non-recyclable items are also buried at landfills. According to the 2011 *White Paper on the Environment and Sustainable Society*, the total amount of general waste disposed in Japan during the year 2009 reached 46.25 million tons, which equates to nearly 1 kilogram (kg) of waste per capita per day. Of this amount, 79.1% was either incinerated upon collection or put through some sort of intermediary processing (e.g. being broken down and sorted). Another 20.5% was collected by recycling businesses, while another 11% (or 5.07 million tons) was buried in landfills.

Comparatively, the total amount of industrial waste for the year 2008 amounted to 43.66 million tons. Of this, 54% was recycled, and 42% was reduced. The 16.4 million tons that were buried at landfills made up a mere 4% of total waste, as recycling and other intermediary processes served to curtail the landfill waste loads.

While these examples illustrate how recycling reduces final waste amounts, diminishing landfill space still exists as a pressing, behind-the-scenes issue. Japan is a small nation with minimal space for final waste-processing lands. Further still, local residents in mountainous areas often object to establishing waste sites nearby with fears that they would inevitably yield hazardous contamination. Thus, to date, no new landfills have been created.

As of the year 2007, the remaining space appropriate for general waste dumping grounds in Japan could accommodate general waste for approximately 15.7 years, while space appropriate for industrial waste was predicted to last for less than 8.5 years. In other words, if things continue as they are, Japan will run out of space for waste in around a decade. Therefore, curbing waste production, reducing the need for landfills via recycling, and extending the lifespan of current landfills have made the top of the political agenda.

Recycling and sustainable society

Due to the challenges described above, multiple laws related to recycling and a sustainable society were passed between the late 1990s and early 2000s (see Table 1.2.1). First, in the year 1995, the *Law for Promotion of Sorted Collection and Recycling of Containers and Packaging* was passed, aiming to promote the collection and reuse of plastic and glass bottles. After that, laws targeting specific commodities, such as the *Home Appliance Recycling Law*, the *Construction Recycling Law*, the *Food Recycling Law*, the *End-of-Life Vehicle Recycling Law*, and various other recycling laws were enacted.

In turn, the *Basic Law for Establishing the Recycling Based Society* was implemented in the year 2000 as a standard for aforementioned recycling-related laws. According to this law, a *sustainable society* is defined as one that limits waste production, and through sustainable use and proper disposal of resources, it reduces environmental strain to the greatest extent possible. The order of precedence for a waste management protocol is as follows:

1. Limit Production (Reduce),
2. Reuse,
3. Material Recycle,
4. Thermal Recycle, and
5. Proper Disposal.

According to these prime directives, the priority is placed on curtailing waste creation, alluding to the fact that recycling alone cannot solve waste issues. Indeed, the *Basic Law for Establishing the Recycling Based Society* is a legal standard for various individual recycling laws, yet it happens to have been established after other recycling laws. While it is legally compelling and possesses all the legitimizing factors of an overarching law, when it comes to enforcing the transition to sustainable society it is not always consistent with all of the other laws.

Let's take a closer look at the *Law for Promotion of Sorted Collection and Recycling of Containers and Packaging*, which was the first of the previously mentioned laws to be established. This law focuses on plastic and glass bottles, paper beverage cartons, aluminum cans, and other such containers. These containers make up around 61% of all general waste (and 22% of the total weight of general waste), making them an important concern in waste-processing deliberations. Before this legislation was put

Table 1.2.1 Laws related to recycling and a sustainable society

Year enacted	Title of law	Contents
June, 1995	Law for Promotion of Sorted Collection and Recycling of Containers and Packaging	Recycling plastic bottles and other containers
June, 1998	Home Appliance Recycling Law	Recycling used home electric appliances
May, 2000	Construction Waste Recycling Law	Recycling construction waste
June, 2000	Food Recycling Law	Minimizing and recycling food waste
May, 2000	Law for the Promotion of Utilization of Recyclable Resources	Promote reduce, reuse, recycle
May, 2000	Law Concerning the Promotion of the Procurement of Eco-Friendly Goods and Services by the State and Others	Procurement of environmentally conscious goods by public institutions
June, 2000	The Basic Law for Establishing the Recycling-based Society	Indicate a target sustainable society
May, 2000	Law Concerning Recycling Measures of End-of-Life Vehicles	Recycling used automobiles

in place, container disposal was managed at the municipal level (i.e. cities, towns, and villages). However, the law dictated that consumers be responsible for separating trash as they disposed of it, that municipalities collect properly sorted waste, and businesses reuse disposed materials when possible. Thereby, a three-entity division of obligatory roles came into existence (see Figure 1.2.2).

Furthermore, upon the execution of this very law, plastic bottle collection rates rapidly increased (as depicted in Figure 1.2.3). In the year 1995, collection rates were no more than 1.8%, but by the year 2009, they had risen to 50.9%. Unfortunately, as Japan also witnessed a sharp rise in production at that time, the collection and management costs borne by municipalities also skyrocketed. Sterilized, deconstructed plastic bottles are processed into fibers or sheets and reused. While "bottle-to-bottle" technologies for creating new plastic bottles out of used bottles are being developed, they are currently quite expensive, preventing their widespread diffusion.

In order for plastic bottle recycling systems to be considered viable, recycling companies must earn profits through the sale of goods made out of recycled goods on the market. However, in many cases, it costs more to produce goods with recycled plastics than it does to create new fibers directly from crude oil. Thus, the fact that recycled goods are not competitive on the market would leave recycling companies in an unfavorable position. Moreover, there are some businesses that are stuck without sufficient used plastic-bottle resources, as much of what is collected is shipped overseas to China and elsewhere. The resource shortages that China faces as its economy grows, along with the steep rise in crude oil prices that were the case as of a few years ago, have both spurred on plastic-bottle resource recycling throughout Japan. The Ministry of the Environment of Japan estimates that nearly 200,000 tons of disposed plastic bottles were shipped overseas in the year 2004. And speaking of waste exports, the *Basel Convention on the Control of Transboundary Movements of Hazardous Wastes and their Disposal* prohibits international waste flows, establishing the premise that waste must be disposed of domestically. Yet because used plastic bottles are sent to China not as waste but as valuable resources, it is difficult to restrict their exportation.

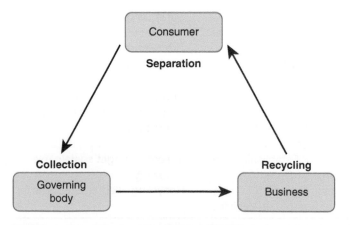

FIGURE 1.2.2 Law for promotion of sorted collection and recycling of containers and packaging

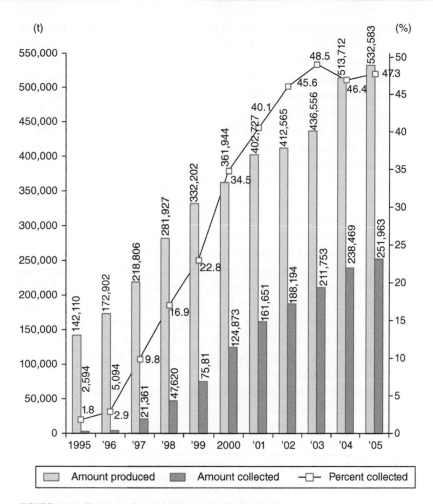

FIGURE 1.2.3 Plastic bottle production and collection levels

Source: Created by the Ministry of the Environment with data from the Plastic Bottle Promotion

Thus, it is clear that while the *Law for Promotion of Sorted Collection and Recycling of Containers and Packaging* has brought about higher reusable resource collection rates, it is quite costly to properly sort and manage what is gathered, effectively impeding proper container recycling and reuse. And even though the law was revised in June of 2006, there are still many aspects beyond what has been said here that must be taken into account before a sustainable society can be attained.

An economic model for recycling

Vast financial resources are required to collect, sort, and reuse materials in recycling schemes. Accordingly, aiming for a sustainable society requires due consideration of economic efficiency in order to cut down on the recycling costs to the greatest extent possible.

Figure 1.2.4 depicts the economics of efficient recycling. As an example, consider a plastic bottle recycling scenario. X_{MAX} represents plastic-bottle disposal amounts in the absence of recycling systems. In this case, if as much as X_1 is recycled, the remaining plastic bottles, amounting to as much as $(X_{MAX} - X_1)$, are discarded as waste. The graph's marginal abatement cost curve indicates the additional cost required per each additional unit (i.e. the marginal cost) of recycled plastic bottles. At X_1, it costs P_1 to increase recycling operations by one unit. When existing recycle levels are low, it is possible to recycle at comparatively low cost. However, as recycling levels rise, considerable labor, facility, and other additional costs involved in collection, sorting, and management begin to grow. In line with this, the marginal abatement cost curve proceeds upward and to the right. Furthermore, it is evident that when all of the available plastic bottles are recycled, recycling can no longer be increased. This circumstance, represented by X_{MAX} on the graph, is where the marginal abatement cost and the slope of its curve become infinite (and thus, the curve is a vertical line).

Now, assume the price of recycled goods is P^*. When the amount of recycling increases by a single unit from X_1, and moreover, when that single recycled good is sold, revenues increase by as much as P^*. On the other hand, the cost increases by as much as P_1, so the remaining balance amounting to as much as $(P^* - P_1)$ is generated as profit for recycling businesses. In other words, to the extent that profits also increase as recycling levels increase, there is adequate incentive for recycling businesses to undertake recycling operations.

Next, consider the scenario where recycling levels are X_2. Here, per each additional unit of recycling, sales revenue increases by as much as P^*, yet costs increase by as much as P_2. Therefore, costs exceed revenues, and the more recycling performed, the

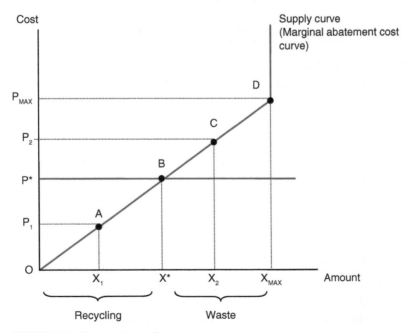

FIGURE 1.2.4 Waste and recycling

larger the recycling company's deficit becomes. In this scenario, it is in the best interest of recycling businesses to decrease the amount it recycles.

The most efficient recycle level is found at X^* on the graph, where the cost of recycled goods and the marginal abatement cost are equivalent. At this point, should the business choose to decrease or increase its recycling levels, it would lose out on profits. If, while ignoring economic efficiency, all disposed plastic bottles were recycled, business operations would become extremely costly. Therefore, regardless of whether or not goods are recycled, if they are not then sold and connected to earning profits, then the recycling firm would experience a deficit.

SUMMARY

As mass consumption societies produce enormous amounts of waste, dwindling landfill and waste-processing space is a pressing issue. In response to this, a number of laws regarding recycling and a sustainable society were put into place. Yet recycling is still quite costly, and recycled goods must be sold after they are processed so that recycle businesses can earn profits. Thus, in order to realize a resource efficient, sustainable society, it is imperative to apply the tenets of economic efficiency in order to make restricting waste output (and, in turn, recycling) as cheap as possible.

REVIEW PROBLEMS

1. The price of used paper was quite high during the oil shocks in the 1970s, but after that it sharply fell. In another scenario, recycling magazine paper could cost close to nil, while in other cases, one would pay to have used paper collected through inverse onerous contracts. Explain why and how the price of used paper can have such ups and downs in this way.

2. Consider the following argument and explain whether or not it is valid, and why:

   "Recycling promotes thorough collection and separation of garbage, so it should be implemented as a means of reducing waste."

3. Consider the following argument and explain whether or not it is valid, and why:

   "In order to curb waste, there should be fees for garbage disposal."

LEARNING POINT: AN ECONOMIC EXPERIMENT WITH WASTE PROBLEMS

The following is an opportunity to witness the nature of waste problems through a simple economic experiment. It can serve as a means of investigating economic structures by reproducing conditions similar to society within the classroom or research lab and of observing the economic behavior of afflicted people.

Take, for example, the question of whether or not plastic items other than plastic bottles should be meticulously separated and recycled (see Figure 1.2.5). As dry-cell batteries contain cadmium and lead, pre-processed trash contains a mixture of toxic components that must be separated before burial. Assume this to be the first round of waste separation. Next, the second round of separation should involve removing plastic bottles and other plastic waste. Assuming there are markings attached to plastic bottles and other plastic goods to distinguish which category they fit into, separating them is a comparatively simple task.

Since there are so many different types of plastic, the U.S. Society of Plastics Industry (SPI) came up with the markings depicted in Figure 1.2.5 to indicate the specific material properties of various plastic items. With this in mind, the next investigation concerns how plastics are marked, separated, and collected based upon their material composition.

Although unrealistic, this economic experiment allows for this system to be tested in the most basic sense. Figure 1.2.6 lays out the waste reduction experimental

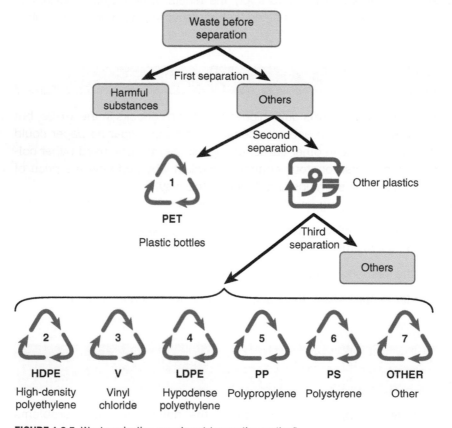

FIGURE 1.2.5 Waste reduction experiment (separation method)

methodology. First, create a 53-box chart as shown in the diagram. Next, write the word "toxic" in one of the boxes. This means that the box contains harmful toxic waste. Finally, randomly fill in the remaining boxes with the numbers 1–13, and repeat this final step four times until all of the boxes are filled in. Every number shall represent a particular type of waste (see Figure 1.2.6).

It takes time to reduce waste costs through proper separation. In the first separation stage, fill in the "toxic" square, and calculate the cost using a rate of ¥1 per second. Record this in the "separation stage 1" box. Next, for separation stage 2,

FIGURE 1.2.6 Waste reduction experiment

Waste situation

3	7	11	5	5	4	7	12
8	Hazardous	3	13	2	11	1	8
6	12	6	1	9	13	12	7
12	9	10	10	13	9	5	6
2	10	13	4	2	5	9	10
8	1	6	11	3	1	7	2
4	3	8	11	4			

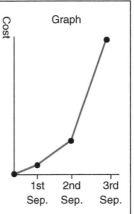

Graph

Cost — 1st Sep. 2nd Sep. 3rd Sep.

Abatement cost (¥1/second reduction time)

1st Separation		2nd Separation		3rd Separation	

Experiment procedure

1. As described above, construct a 53-boxed chart. When making an 8 by 7 chart, cross out three boxes as shown in the figure above.
2. Label one box as "Hazardous"
3. Put the numbers 1 through 13 into random boxes.
4. Repeat step 3 four times until every box is numbered.
5. Separation 1: Shade in the box labeled "Hazardous" while recording the time it takes. When finished, record this time in the "1st Separation" abatement cost section.
6. Separation 2: Locate the 4 plastic bottle boxes labeled "1" and shade them in. Record the time it takes in the "2nd Separation" abatement cost section.
7. Separation 3: Locate the 4 plastic boxes labeled "2" and shade them in. Repeat this process for the 4 boxes labeled "3." In the same manner, sequentially locate and shade in all numbered boxes through the 4 labeled "7" and record the time it takes in the "3rd Separation" abatement cost section. If this is not completed within 60 seconds, mark 60 seconds.

Record the separation numbers and abatement costs on a graph. (This experiment can also be performed with cards. In such a case, a Joker card represents the "Harmful" box, cards 1–13 correspond with the same numbers above, and instead of shading in boxes, the cards ought to be sequentially ordered to perform the analysis.

locate and fill in the four boxes marked with the number "1," all of which signify plastic bottles. In the same manner as before, calculate the reduction cost using a ¥1 per second rate, and record the results in the "separation stage 2" box. For stage 3, the remaining plastics are to be separated and categorized based upon their material makeup. First, locate and fill in the four boxes marked with the number "2," which denote high density polyethylene. Next, find the four "3" marked, vinyl chloride boxes and shade them in. Repeat this process until all of the remaining boxes through those marked with the number "7" are shaded in sequential order, and record the reduction cost using the same method as above in the box labeled "separation stage 3." Finally, construct a first quadrant graph, labeling the separation stages along the horizontal axis, and mark the cost measurements along the vertical axis. Then, plot the results of the experiment and construct a marginal abatement cost curve that proceeds up and to the right, as shown in Figure 1.2.6.

Through this experiment, it is clear that the three stages for separating plastic goods based on their material makeup are quite costly. It follows, then, that the cheaper recycled goods are, the less economically efficient it is to recycle them. While this type of oversimplified experiment does not necessarily lead to accurate predictions in the real world, it does elucidate the inherent costs of recycling that must not be overlooked or underestimated in a sustainable society.

The economic experiment introduced here can also be performed online.

Section 1.3: Climate change

What is climate change?

Climate change is another environmental issue that is intricately linked to our lifestyles. Because carbon dioxide (CO_2) and other climate change-inducing *greenhouse gases* are emitted when we use energy sources such as gas and electricity, in order to prevent climate change, we must reconsider our dependence on fossil fuel energy sources.

The difference between the amount of solar energy that reaches the Earth and the amount that is released back into space determines global temperatures. However, CO_2, methane, and other greenhouse gases absorb the heat reflected of the surface of the Earth. Therefore, as greenhouse gas concentrations in the atmosphere accumulate, less heat is released into space and global temperatures rise.

Gases such as CO_2 and methane are produced through various human undertakings, from burning oil and other fossil fuels to modifying natural lands to suit human needs. Figure 1.3.1 illustrates the changes in atmospheric concentrations of CO_2 from 10,000 years ago through the present. The pre-Industrial Revolution era was marked by stable, low-level alterations of CO_2 concentrations, yet subsequent eras

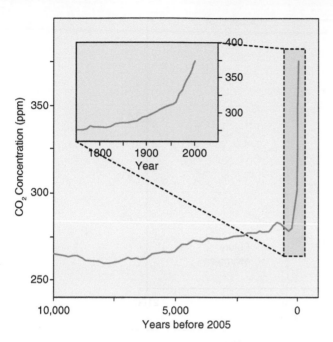

FIGURE 1.3.1 Carbon dioxide concentration trends

Source: *Contribution of Working Group 1 to the Fourth Assessment Report of the IPCC*, February, 2007

of rapid economic development were marked by a rapid escalation in fossil fuel use and corresponding influxes in atmospheric CO_2 concentrations.

What type of changes will occur in the Earth's climate should the atmospheric concentrations of CO_2 and other greenhouse gases continue to accrue? According to the *Contribution of Working Group 1 to the Fourth Assessment Report*, published in February of the year 2007 by an international organization of government representatives and technical experts known as the Intergovernmental Panel on Climate Change (*IPCC*), average surface temperatures on Earth are predicted to rise somewhere between 1.4 and 4° Celsius (C) compared to 1990-year levels by the year 2100. See Figure 1.3.1. Figure 1.3.2 portrays the results of the estimates made by the IPCC. Were each nation to pursue its unilateral economic growth aspirations, CO_2 emissions would surge drastically, and atmospheric temperatures would consequently climb by 3.4°C by the year 2100. On the other hand, temperature rises could be restrained to within 1.8°C through energy conservation and technological development. Furthermore, if CO_2 concentrations could be fixed to year 2000 levels, it would be possible to limit atmospheric temperature increases to within a mere 0.6°C threshold. Overall, then, these figures suggest that the state of the Earth 100 years from now is largely dependent upon two major points: whether or not the world's economic development continues down its current path, and whether or not the international community can cooperate in adopting climate change policies.

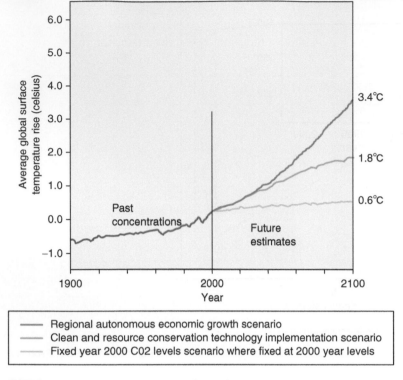

FIGURE 1.3.2 Future global temperature fluctuation estimates

Source: *Contribution of Working Group 1 to the Fourth Assessment Report of the IPCC*, February, 2007

The impact that climate change will have on society

As climate change intensifies, how will society be impacted? First of all, there is the issue of rising sea levels. Some predict overall sea-level rises as rising global temperatures will cause thermal expansion of sea water and glacial melting. The IPCC estimates that climate change will cause sea levels to rise by 0.18 to 0.59 meters by the year 2100. Because of this, many island countries in Oceania at low positions above sea level face the threat of losing land to complete inundation as climate change triggers rising average sea levels. This is particularly true for the South Pacific Island of Tuvalu. Some experts anticipate that once its territories are largely submerged under rising sea levels, many of its displaced citizens will emigrate to foreign countries as environmental refugees. Accordingly, regional neighbors, such as New Zealand, are devising immigration policies well in advance.

Other salient issues are the droughts and floods that will undoubtedly adversely affect societies as precipitation levels and patterns alter. As precipitation changes will vary regionally, some areas are expected to suffer severe droughts, while other areas will likely experience excessive precipitation and disastrous flooding. The IPCC estimates that in addition to the hundreds of millions of people who will be afflicted by

TABLE 1.3.1 Climate change damage predictions

Water resources	Billions of people to face water shortages
Ecosystem	Numerous wildlife species will go extinct
Food production	Harvest yields will fluctuate regionally
Coastal areas	Large damages brought about by flooding and extreme storms; many coastal wetlands will be lost
Health hazards	Tropical region-based contagions will spread; health hazards related to heat waves and flooding will increase

Source: *Contribution of Working Group 2 to the Fourth Assessment Report of the IPCC*, April, 2007

droughts, the devastation wrought by raging storms along coastal areas and other natural disasters will only grow worse, meaning millions of people will suffer disaster-inflicted damages every year.

Next, climate change will pose dire ecological repercussions. As major environmental changes take effect, many living creatures may very well be unable to adapt to their altered natural habitats. A great number of these species will reach or surpass the verge of extinction. According to the IPCC, a rise in atmospheric temperatures by a mere 1°C could lead to the extinction of 30% of the wildlife species on Earth.

Moreover, climate change will drastically affect agricultural productivity. As atmospheric temperatures rise, growing conventional agricultural products stands to become more challenging than before, and water shortages will only exacerbate substantially fewer harvest yields. However, as the impact that climate change will have on agriculture is anticipated to vary with location, some regions may experience diminishing crop outputs while others may actually experience greater yields.

Finally, climate change is anticipated to affect the health and lives of people. As heat waves, floods, and other irregular weather patterns occur more frequently, many multitudes of people will risk losing their lives. Furthermore, malaria and other infectious diseases that had previously only broken out in tropical regions could potentially spread to other non-tropical regions.

In all of the aforementioned ways, climate change is expected to have tremendous impacts on both human society and economic activities, so there is an undeniable exigency to ratify climate change policies at the international level.

Climate change and economic models

At this point, economic models can be employed in consideration of the mechanisms that underlie climate change issues that result from economic growth. Economic development entails greater consumption of oil and other fossil fuel energy sources, leading to greater CO_2 emissions upon fuel combustion and even riskier climate change dangers. The amount of fossil fuel energy consumed is determined by its supply/demand balance.

Figure 1.3.3 displays the demand for fossil fuel energy. Assume that when the price of oil is P_1, as much as X_1 oil is consumed, and likewise, when the price of oil is P_2, as

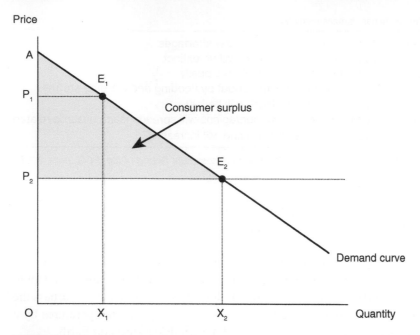

FIGURE 1.3.3 Fossil fuel energy demand curve

much as X_2 oil is consumed. The demand curve on the graph depicts this relationship between price and consumption. In general, oil consumption is assumed to increase to the extent that it is cheap, and conversely, consumers restrict their oil use when prices rise. Thus, the demand curve proceeds downward and to the right, as shown on the graph.

Assume that the current price is P_2. In this scenario, when consumption levels are X_1, the consumer would likely purchase as much as P_1. However, since the actual price is P_2, the consumer earns profits equal to $(P_2 - P_1)$. Similarly, in other consumption scenarios, the area of the gap between the actual price and the demand curve amounts to profits pocketed by the consumer. That area, represented in this case by $\triangle AP_2E_2$, is known as the *consumer surplus*.

Figure 1.3.4 shows the supply curve for fossil fuel energy. Consider, for example, that when the price of oil is P_1, then as much as X_1 oil is produced, and when the price is P_2, then as much as X_2 is produced. Here again, the supply curve can be derived on a graph by expressing this relationship between price and production levels. Generally speaking, since production levels grow to the extent that prices are high, the supply curve is one that proceeds up and to the right, as shown on the graph.

The supply curve could otherwise be called the marginal cost curve. Marginal costs represent the added costs accrued per each additional unit of production. Producers determine production levels with the aim of maximizing profits, so they will produce up to levels where price and marginal cost are equivalent. For example, assume that the current price is P_2. In this scenario, assuming as much as X_1 is produced, were production to increase by a single unit, income would increase by as much as P_2.

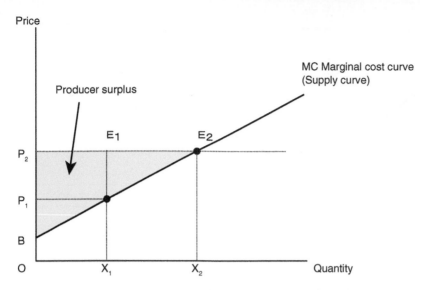

FIGURE 1.3.4 Fossil fuel energy supply curve

However, when the marginal cost is P_1, overall costs increase by as much as P_1. Therefore, the price difference of as much as $(P_2 - P_1)$ that would arise per each additional unit of production would amount to profits for the producer. In this manner, when price surpasses marginal cost, there is potential to earn greater profits, so producers are inclined to increase production up to X_2, where price is equivalent to marginal cost. However, when production levels begin to outstrip marginal cost, overall cost increases exceed any additional revenue earned through production increases, so production will not be expanded to this extent. Accordingly, since producers fix production to levels where price and marginal costs are equivalent, the marginal cost curve becomes the same as the supply curve. Moreover, since the difference between price and marginal cost leads to profits for the producers, the area of the space between the price and the marginal cost curve (represented in this example by ΔBP_2E_2) is known as the *producer surplus*.

The actual price and amount of oil used is determined by the supply-demand balance in oil markets. Figure 1.3.5 displays this balance. Point E, where the supply and demand curves intersect, is where the supply and demand are balanced, and subsequently, where the equilibrium price P^* exists. Here, the *social surplus*, comprised of the sum of both the consumer and producer surpluses, is maximized, representing the most efficient condition. In other words, by adjusting prices through market mechanisms, supply and demand come into balance, and efficient production can be achieved.

At this point, assume that the demand for fossil fuel energy increases alongside economic development. Figure 1.3.6 depicts such a demand increase. When the demand increases, the demand curve shifts from D to D'. In this scenario, the previous equilibrium price P^* is supplanted by F, located at the intersection between the supply curve

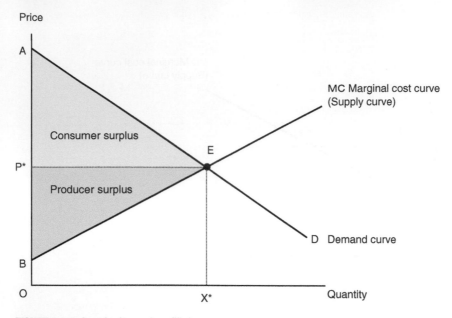

FIGURE 1.3.5 Supply-demand equilibrium

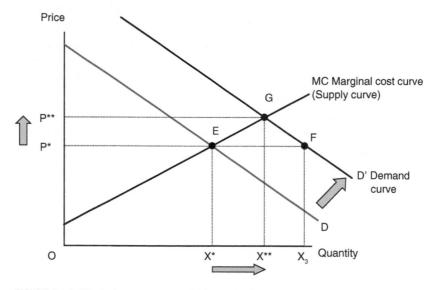

FIGURE 1.3.6 The impacts of increased demand

and the new demand curve. Accordingly the new demand amounts to X_3. However, the producer will only produce at prices equivalent to those found at X^*, and out of this, excess demand is born. The result is that prices increase. Yet when prices increase up to P^{**}, a new intersection point G between the demand and supply curves is formed, bringing supply and demand into balance. Through this process, the amount of energy used through economic development increases from X^* to X^{**} while energy prices increase from P^* to P^{**}.

When use increases, so to do CO_2 emissions, adding to the already enormous risks of climate change. Yet, as demonstrated previously, market mechanisms that do indeed succeed at preserving the balance between fossil fuel energy demand and supply, do not properly account for climate change and its effects. In other words, market mechanisms do not efficiently function with regards to climate change, and solving climate change would be unattainable if the countermeasures were entrusted to the market alone.

LEARNING POINT: PREVENTING CLIMATE CHANGE

In order to prevent climate change, it is imperative to reduce the emissions of the greenhouse gases that cause it. However, since market mechanisms do not effectively function for climate change, even if the market is entrusted, emissions reductions will be achieved. What's more, since climate change takes place on a global scale, single-nation emissions reductions would prove ineffectual, and global level cooperation is needed.

Because of this situation, the United Nations Framework Convention on Climate Change was adopted at the Earth Summit held in 1992. This system established a framework for initiatives related to climate change, and it called for developed nations to reduce CO_2 and other greenhouse gas emission amounts to 1990-year levels by the close of the 1990s. In the year 1997, the Kyoto Protocol was adopted to stipulate the upper limits of greenhouse gas emissions, and Japan set a target of reducing its emissions by 6% compared to 1990-year levels. Furthermore, the Kyoto Protocol approved of three particular processes for reducing emissions at low cost:

1. "Joint Implementation" (JI) for cooperation in emissions reductions among developed countries,
2. "Clean Development Mechanisms" (CDM) for developed countries to contribute funding to emissions reductions in developing countries, and
3. "Emissions Trading" for buying and selling emissions amounts internationally.

However, the United States, as the largest emitting country in the world at the time, chose not to ratify the protocol out of concerns that its parameters would undermine the U.S. economy. Furthermore, the protocol only obliged developed nations to cut back on their emissions, with no specific emissions reduction targets for developing nations that are expected to emit more as their economies grow. Creating an international climate change protocol that includes targets and procedures for both developed and developing nations remains at the heart of ongoing climate change discourse.

SUMMARY

Climate change risks increase in severity as greater fossil fuels are consumed to drive economic development. The balance between the supply and demand of fossil fuels dictates fossil fuel energy consumption. However, these market mechanisms do not necessarily function effectively in the case of climate change, as they fail to account for climate change damage costs. In future, adopting protocols for international climate change policy that include specific emissions reduction targets for developed countries as well as roles for developing countries will be a matter of high concern.

REVIEW PROBLEMS

1. Summarize what could be done in order to prevent climate change.
2. Consider the following argument and explain whether or not it is valid, and why:

   "As economic development leads to increased oil use, oil prices will also rise, causing businesses and consumers to begin to conserve and cut back their oil use. Thus, climate change will not come into effect."

3. The demand function for oil is $x = 4 - p$, and the supply function is $x = p - 2$. P is the price of oil, and x is the amount of oil. With this information, solve the following problems:

   a. Derive the equilibrium price and the amount of oil used at this point.
   b. Through economic development, demand increases to the point where the demand function becomes $x = 8 - p$. Determine how much the amount of oil used increases by at this point.
   c. Calculate the consumer and producer surpluses when demand is maximized.

2

Mechanisms that give rise to environmental problems

Chapter overview

This chapter is devoted to explaining the economic mechanisms that bring about environmental problems. Market mechanisms that fail to efficiently cope with environmental problems, also known as "market failures," have become causal factors of environmental issues. In this chapter, economic models are used to explain this causality.

The first section examines the nature of externalities (i.e. external diseconomies). Water and air contamination, along with other forms of environmental pollution that are ignored by the market, pose considerable health risks to local residents. These phenomena form one type of "negative externality" which fails to consider the impacts of pollution on the environment in the context of market economics, leading to excessive pollution.

The second section discusses issues related to common-pool resources. These include forest and aquatic resources, as well as other natural provisions that are used and managed by people collectively. Were these resources open to unrestricted access by anyone and everyone, so long as they continued to be profitable, they would be used up through exhaustion. Due to this, common-pool resource use must be appropriately managed.

Finally, the last section of this chapter focuses on free-rider problems surrounding public goods. The results of ecosystem preservation and climate change policy affect people everywhere, so in essence, they are public goods. Consequently, this gives rise to the free-rider problem, in which an individual or group may attempt to avoid paying its share of the financial burden of public goods.

Moreover, this chapter includes an economic experiment with environmental issue catalysts in order to shed light on the nature of environmental issue sources and catalysts.

Chapter content

Section 2.1—Environmental contamination is ignored by the market and inflicts harm on local residents. This section provides specific examples of how market mechanisms fail to function efficiently and how these *market failures* cause environmental issues. This is followed up with an economic experiment related to the future impacts of climate change.

Section 2.2—This section opens with a discussion about collectively used, common-pool resources (i.e. the commons). When there is unrestricted access to the commons, there is the risk that they will be overused until they are depleted via a scenario known as the *tragedy of the commons*. This is demonstrated with a discussion of how even fish stocks and other renewable resources could potentially fall victim to excessive exploitation.

Section 2.3—As they stand to impact all of humankind, climate change policies can be viewed as public goods. Unfortunately, climate change policies, like all public goods, are vulnerable to the *free-rider problem* that arises when people avoid paying their share of what is required to maintain the good. This section includes an economic experiment that clarifies complications with public goods.

Section 2.1: Externalities and market failures

Environmental problems and externalities

Market economies lie in the backdrop of environmental issues (see Figure 2.1.1). In market economies, the supply-demand balance determines prices for manufactured goods. Producers commit to production levels that will maximize their profits. At the same time, consumers adjust their consumption levels based on the good prices in order to maximize their own utility. In this way, prices allow for market mechanisms to function in achieving economic efficiency.

However, market mechanisms do not function efficiently when it comes to the environment. Consider, for example, the health problems that arise among residents who

FIGURE 2.1.1 Externalities and environmental problems

live near a smoke (pollution)-emitting factory. In this scenario, the price of factory smoke is not included in the market, so it is ignored and thereby harms local dwellers. Activities that are overlooked by the market and affect other people in this way are called *externalities*. Externalities that benefit third parties are known as *positive externalities* (otherwise known as "external economies"), while those that prove to be detrimental to others are known as *negative externalities* (otherwise known as "external diseconomies"). As such, the previous example in which air pollution from factory smoke imposes adverse health effects on local residents is a negative externality.

Although many environmental problems are negative externalities in the manner mentioned above, further inspection reveals the variety of forms that negative externalities take. Table 2.1.1 provides prime examples of environmental problems related to negative externalities. When factories emit air and water pollution, specific emitters can be pinpointed as the source of pollution, and those who live in the vicinity of or downstream from the emitter are deemed victims. Thus, negative externalities could be brought on by a specific and limited number of actors (in this case, pollution emitters) and cause harm to a specific and limited number of people (in this case, unhealthy local residents).

By contrast, car exhaust-related health problems that arise among residents who live along high-traffic roadways demonstrate a different scenario. While the roadside residents easily could be specified as victims, it is difficult to determine who, out of the large number of drivers who pass through the area, is to blame for the problem. Scenarios involving tropical rainforest destruction present similar ambiguity when attempting to identify culprits. While local residents who convert tropical rainforests into farmland can be singled out as culprits, since the resultant loss of biodiversity as many wildlife species are forced to extinction has global impacts, so identifying the number of people who are adversely affected across the planet becomes quite problematic. And of course, there is the case of climate change, where the actions of a specific number of people in the present could potentially have widespread impact on unspecified numbers of people in the future. It is clear, then, that the term *negative externality* encompasses a broad spectrum of factors, and as such, it is important to determine how culprits and victims can become more inclusive in relevant countermeasures, both spatially and temporally.

Table 2.1.1 Environmental problems and externalities

Issue	Culprits	Victims
Pollution	Few and specified (factories)	Few and specified (local residents)
Automobile exhaust gas	Many and unspecified (numerous drivers)	Few and specified (roadside residents)
Tropical rainforest destruction	Few and specified (residents in tropical regions)	Many and unspecified (people around the world)
Climate change	Many and unspecified (current generation)	Many and unspecified (future generations)

Externalities and market failures

Externalities and market mechanism failures lead to phenomena known as *market failures*. Consider greenhouse gas emissions from an electrical power plant. A common way to generate electricity is to burn oil to produce thermal power. However, one of the main byproducts of this process is the climate change-inducing CO_2, and large loads of it are emitted into the atmosphere. While there are various existing power generation methods (e.g. nuclear power, wind power, etc.), for simplicity, the focus will be on thermal power generation.

Figure 2.1.2 displays market equilibrium in an electricity market. The supply-demand balance of electricity determines the market price. Curve *D* represents the demand curve for electricity. The cheaper electricity is, the more it is consumed, forming a down-rightward demand curve. In the same figure, the *MPC* curve depicts the marginal private costs. As such, it depicts the additional costs required to increase the electricity supply by a single additional unit, and it is the same as the supply curve for electricity. While the *MPC* curve does incorporate the construction and maintenance costs involved in supplying electricity, it does not include the impacts of climate change. Point *E*, where the supply and demand curves intersect, is where supply equals demand and market equilibrium is realized. Here, the market price is P^E, while the electricity production and consumption levels are X^E. As shown in Figure 2.1.2, the consumer and producer surpluses combine to form a total surplus equal to the area of $\triangle ABE$.

Next, consider the impact of climate change caused by thermal power generation. Since the damages inflicted by climate change are anticipated to severely impact

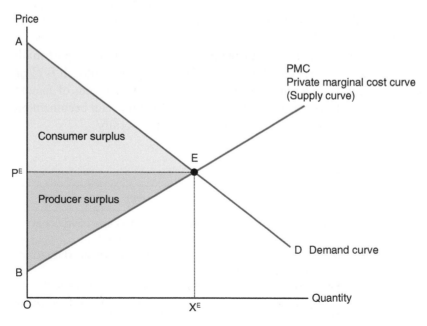

FIGURE 2.1.2 Market equilibrium

future generations, thermal power generation (as a catalyst of climate change) could thus be considered to bring about negative externalities for future generations. Therefore, it is important to include both immediate costs (facility construction, etc.) as well as future damage costs (externalities) when evaluating electricity supply.

Figure 2.1.3 portrays an electricity market that properly accounts for externalities. First, look at the bottom graph. It depicts the climate change damage as a consequence of thermal power generation. The *MEC* curve represents marginal externality costs, indicating the future additional damage costs of climate change per additional unit of electricity supplied. Climate change impacts grow more severe as more electricity is supplied. Thus, the *MEC* curve proceeds upward and to the right.

The top graph, however, includes an *MSC* curve, which adds marginal externality costs to marginal private costs, depicting the marginal social cost. For instance, when electricity supply levels are X^E, marginal private costs reach $X^E E$ on the top graph, while marginal externality costs on the bottom graph become $X^E H$ (equivalent to *EF*). When the two are combined, they form the marginal social cost $X^E F (= X^E E + X^E H)$. The *MSC* curve includes not only such private costs as power plant construction, but also such externality costs as climate change effects. Accordingly, when climate change impacts are considered, the *MSC* curve and the demand curve *D* intersect at *E*, the socially optimal point. Here, the price of electricity is set to the socially optimal price P^*, and electricity supply is set to the socially optimal amount X^*. Since market

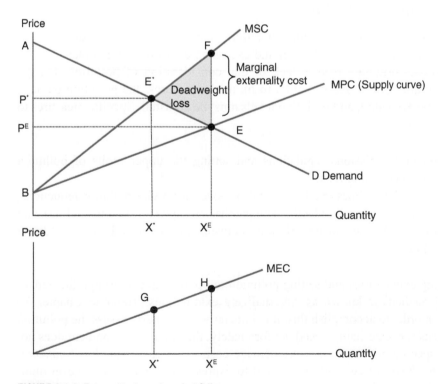

FIGURE 2.1.3 Externality-based market failure

mechanisms determined the electricity supply levels at market equilibrium to be X^E, electricity up to the X^*X^E maximum are consumed. However, when negative externalities exist in markets, market equilibrium does not ascribe value to externalities, resulting in overproduction.

Figure 2.1.3 can be used to grasp how overall surpluses change with the presence of externalities. At the socially optimal point E^* where externalities are properly reflected, the consumer surplus is equivalent to $\triangle AP^*E^*$, the consumer surplus is equivalent to $\triangle BP^*E^*$, and the overall surplus amounts to $\triangle ABE^*$.

On the other hand, at market equilibrium point E when externalities are ignored, the consumer surplus is equivalent to $\triangle AP^E E$, while the producer surplus is equivalent to $\triangle BP^E E$. Since these combine to form $\triangle ABE$, upon first glance, it may appear as if the surplus only increased by $\triangle BEE^*$. However, externality costs equivalent to $\triangle BEF$ brought about by climate change must be subtracted from the surplus. Upon doing this, compared to what is left when market equilibrium is socially optimal, surpluses of as much as $\triangle E^*EF$ are lost. This loss of $\triangle E^*EF$, or any other loss that results when externalities are ignored, is known as a *deadweight loss*.

Therefore, when negative externalities are present, market failures arise as market mechanisms cease to function efficiently and overproduction leads to forfeited surplus.

Internalizing externalities

When externalities give rise to market failures, entrusting their resolution to the market alone leads to overproduction and excessive environmental contamination. Market failures originate from a lack of due consideration of pollution damage costs (externality costs). Accordingly, to fix market failures, the pollution damage costs must be evaluated, and pollution levels must be adjusted. Specific measures to achieve this include:

1. Implementing emissions regulations and setting the upper limits of pollution (environmental regulations),
2. Imposing pollution fines (environmental taxes) to incentivize pollution reductions,
3. Establishing emissions production and purchasing rights (emissions trading), and
4. Curtailing pollution via negotiations between polluters and victims (direct negotiations).

Reviewing externalities and setting pollution to optimal levels through the above-mentioned methods is known as "internalizing externalities" (refer to Chapter 3). However, in order to accomplish this, it is imperative to properly assess the pollution damage costs (i.e. externality costs). Unfortunately, the natural environment does not possess a specific price, so assessing damage costs is no easy feat. *Environmental valuation methods* are currently employed to express environmental value in monetary units (refer to Chapter 5).

LEARNING POINT: AN ECONOMIC EXPERIMENT WITH EXTERNALITIES

The following simple economic experiment provides a closer look at the nature of externalities. Figure 2.1.4 displays the experimental methodology. Players are grouped as "current generation" and "future generation." If only one person is performing this experiment, then they should carry out both roles consecutively. If there are multiple people, then the roles should be separated.

The experiment proceeds as follows. To start, create a five by five grid. In the first stage, the current generation pollutes the environment as it manufactures goods. Profits increase to the extent that production increases, but at the same time, pollution also increases. In this experiment, placing the letter "O" on grid spaces represents pollution. For five seconds, the current generation participant should fill as many Os into the empty grid spaces as possible. One O amounts to ¥100,000,000 in profits. Accordingly, the current generation can increase profits to the extent that they fill in Os within the given timeframe. After 5 seconds pass, calculate the current generation's profits. In the example shown, 15 Os are filled in, so the current generation's profits amount to ¥1,500,000,000.

The next step is for future generations to clean up pollution. This is portrayed in this experiment by fully shading in grid squares. For 5 seconds, the future generation participant should shade in as many O-marked grid spaces as possible. After 5 seconds, note the amount of pollution that could not be eradicated. The remaining grid squares with O marks each represent ¥1,000,000,000 in damage

①	②	③	④	⑤
⑥	⑦	⑧	⑨	⑩
⑪	⑫	⑬	⑭	⑮
16	17	18	19	20
21	22	23	24	25

Pollution level (1)	15 tons
Current generation profits (2)	¥1.5 billion
Amount removed (3)	5 tons
Amount of remaining contamination (1)–(3)	10 tons
Damage cost to future generations (4)	¥10 billion

Experimental procedure

1. Construct a 5×5 box chart.
2. Current generation: Draw circles in consecutive squares for 5 seconds. The number of circles drawn represents the level of contamination (1 circle is worth ¥100 million in profits)
3. Future generation: Pollution removal. Shade in the circled boxes for 5 seconds.
4. Only the remaining circled boxes yield damages. One circle amounts to ¥1 billion in damage.
5. Calculate the total profits by subtracting the future generation's damages from the current generation's profits.

FIGURE 2.1.4 An economic experiment with externalities

costs. In the example given here, only five squares were completely shaded, so ten remain untouched, meaning the future damage costs are calculated as ten squares × ¥1,000,000,000 (= ¥10,000,000,000). As can be seen here, the future generation must shade in as many contaminated grid squares as possible within the allotted 5 seconds.

Finally, net profits can be calculated by subtracting future damage costs from current profits. In the example, current ¥1,500,000,000 profits − future damage costs of ¥10,000,000,000 = −¥8,500,000,000. This experiment models how pollution in the present gives rise to future externalities. Generally speaking, it is easy to pollute, yet it is quite costly to clean it up. The comparative tedium of marking Os for pollution and shading in grid boxes for pollution cleanup alludes to how much more difficult the latter is over the former. Ultimately, purely profit-driven, excessive pollution in the present leads to remarkable damages in the future.

SUMMARY

Pollution is inherently a negative externality in that it is overlooked by the market and it deals damages to third parties in the absence of properly functioning market mechanisms. To this end, pollution becomes excessive and unchecked when the market is the only regulatory framework for environmental stewardship. It is therefore imperative to evaluate the externality costs of pollution and suppress them appropriately.

REVIEW PROBLEMS

1. Assume x hectares (ha) of tropical rainforest is being converted into farmland. The conversion cost is represented by the function $2x^2 + 2x$. Also assume that the market price of farmland is ¥p/ha, and the demand function for farmland is $x = 7 - 0.5p$.

    a. Determine the agricultural land conversion area and market price at market equilibrium.
    b. Assume damages amounting to ¥$6x$ are incurred when x ha of tropical forest are lost to agricultural conversion. Determine the socially optimal farmland conversion area that includes the impact of tropical rainforest losses under such circumstances.
    c. Calculate the deadweight loss.

2. Explain how pollution issues inherently involve negative externalities.
3. Investigate what kind of policy would be needed to reduce damages inflicted on future generations based on the economic experiment with externalities shown in Figure 2.1.4 of the **Learning point.**

Section 2.2: The use and management of common-pool resources

The tragedy of the commons

The air, forests, rivers, oceans, landscape, and other features of the natural environment are resources that everyone uses and manages jointly, and they are not exclusively possessed by any single individual or entity. Resources such as these are known as *the commons* or *common-pool resources*. The atmosphere is a prime example. Unfortunately, factory gas emissions, automobile exhaust fumes, and a host of other scenarios illustrate how many entities regard the atmosphere as a free, emissions dumping zone. Since it is difficult for any sole entity to claim exclusive rights to use the atmosphere, a great number of factories and individuals all use it at their convenience.

As the commons are natural resources that anyone can use freely, were they improperly managed, they could be excessively used to the point of exhaustion. Take, for example, an open pasture that serves as grazing land for cattle. In contrast to grasslands that are managed by exclusive owners who have it in their best interest to limit their livestock numbers in order to prevent over-grazing, grasslands open to unrestricted public use may be overgrazed by one party even if another party decides to restrict their cattle's consumption. Since those who choose to refrain from excessive use would ultimately miss out on profit-making opportunities, most would end up allowing their cattle to graze as much as possible. This thereby represents a scenario where actors, motivated only by personal gain, would allow their cattle stock and consumption to reach levels that would completely deplete the grassland. This type of phenomenon, was first depicted in an essay by Garrett Hardin entitled *The Tragedy of the Commons* in a 1968 edition of the *Science* academic journal.

Figure 2.2.1 depicts the tragedy of the commons. Refer back to the cattle-grazing example from before. This time, assume that ranchers can increase cattle numbers by any amount, and the costs of these aggrandizements are negligible. However, greater livestock levels lead to meadow losses, which is a detriment to society. The horizontal axis measures the cattle count, while the vertical axis measures cost. The marginal benefit in this case is the further profit earned by ranchers per each additional head of cattle. There is also a marginal cost, which in this case is the additional cost of damages dealt to grasslands per each additional head of cattle. The socially optimal number of cattle is realized at point X_1, where the marginal cost and marginal benefit are equivalent. Here, the rancher benefits amount to $A + B + C$, while the damage costs are equal to C and the overall social benefits amount to $A + B$.

However, even at optimal point X_1, should ranchers increase their cattle numbers, their profits will also increase to levels greater than current OP amounts. Even if a given rancher limits his cattle numbers, others aiming for profits equal to OP will increase their cattle numbers. The result is that ranchers constantly scale up their herds so long as they can accrue greater profits. When marginal benefits bottom out,

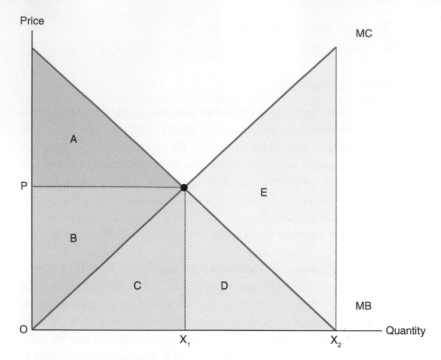

FIGURE 2.2.1 The tragedy of the commons

there is no profit to be gained through greater cattle numbers. This point, shown at X_2 on the graph, represents the greatest cattle count that farmers will pursue. Here, rancher benefits are equal to $A + B + C + D$, but due to grassland loss, damages are worth $C + D + E$, and overall social benefits equate to $A + B - E$. In other words, when ranchers think only of their own profits, they will increase cattle numbers to a point where grasslands are excessively used, and in the absence of herd regulations, society will suffer damages equivalent to E.

In this way, the tragedy of the commons occurs when multiple parties use the commons in an unrestricted, excessive fashion, leading to their complete exhaustion. This trend is prevalent throughout a host of environmental issues, including thinning forests, diminishing fisheries, degraded water quality, and land contamination. Furthermore, as the natural environment is intrinsically a global, jointly used common resource, environmental issues like climate change and tropical rainforest deforestation can be thought of as examples of the tragedy of the commons.

What is needed to preserve the commons?

What, specifically, is needed to prevent the tragedy of the commons? The fact that the commons can be freely used by the masses (e.g. their open access) is the underlying cause of the tragedy of the commons. Therefore, usage restrictions ought to be implemented to prevent their overuse.

The first potential method to achieve this end is to privatize the commons. For example, in the grasslands case, dividing previously openly accessible land into areas that can be used only by designated parties would incentivize users to limit their herd numbers to prevent grassland depletion. Another potential method is to make the commons public. Should national or local governments manage the commons, they could employ various policy controls to prevent excessive use.

However, there are still many challenging aspects to privatizing the commons that should not go unrecognized. For instance, it would be a technological marvel to divide the atmosphere into areas for use only by a select number of entities. Similarly, divvying and privatizing an ecosystem, with its plentiful and interconnected forests, rivers, wildlife, and other components, would be no simple feat.

Moreover, while it is technologically feasible to make certain commons private or public, there are also enormous expenses involved that render them difficult to execute. For example, if frequent tourism at a national park led to litter or other acts that undermine the quality of the natural system, installing entrance gates and restricting visitor numbers could be effective protection measures. In the U.S., many national parks have only a few entryways, so installing gates is feasible. However, in Japan, there are often private lands within by public parks, and many public parks have multiple roads leading into them. Compared to cases in the U.S., erecting park gates at every entrance in Japanese parks would be quite costly, which is why many parks in Japan remain open to free access.

Of course, there are also scenarios where nationalized, relatively protected land resources can be the subjects of improper,excessive use. For instance, in tropical regions, many rainforests are nationalized and forest felling is restricted. However, as it is prohibitively expensive to constantly watch over and safeguard these vast forests, it is difficult to crack down on illegal slash-and-burn farming and forest-felling techniques. As such, preserving the commons in these cases is much easier said than done.

Yet there are also many examples of multiple parties using the commons in a manner that did not lead to their depletion and, in fact, allowed for their preservation. One example of this involves the formerly commonly managed forests throughout Japan. The management and upkeep responsibilities of these forests were shared by local residents. On a daily basis, locals would collect mushrooms, firewood, and other necessary forest resources while adhering to a set of resource-preservation regulations. One of these regulations was to limit the amount of a given resource that could be procured per day. If, for any reason, the local conventions were ignored, rule breakers were restricted from forest access, among other punishments. These types of regulations encourage the appropriate use and management of natural resources, serving as a check on resource depletion and a barrier to the tragedy of the commons.

However, at the start of Japan's post-World War II period of rapid economic growth, electricity and gas fuels diffused even to rural districts. At once, collecting forest fuel sources became less important to the lives of rural people, especially as rural populations proceeded to decline as citizens flocked to metropolitan areas. The commonly managed forests thus lacked the management they needed to thrive, leading to their ever-apparent decline in quality. In other words, the deterioration of the forests was not instigated by resource exhaustion à la tragedy of the commons, but rather, it came about due to decreasing population levels, less dependency on woodland fuel supplies, and other changes among local communities.

Similar examples, where natural resources are kept intact by human conventions over multiple generations, can be found all over the world. In all these cases, environmental regulations with clear usage rules and enforceable punishments for violators, and local governance, were indispensable to the preservation and management of common resources.

Renewable resources and open access

Natural resources can be classified into *non-renewable resources* that cannot be replenished after exhaustion, and *renewable resources* that can be replenished. For example, petroleum is a non-renewable resource. Due to its finite subterranean reserves, petroleum resources are non-renewable. Contrarily, wood, water, agricultural crops, fish stocks, and many other resources found in the natural environment are renewable resources. For instance, even after large-scale lumbering, timber resources can return to their previous level of abundance and quality should they be properly managed alongside reforestation procedures. Naturally, if forests resources are harvested without proper restrictions, management practices, and reforestation efforts in effect, they will more than likely be depleted before they can replenish. Thus, lumber restrictions and post-lumber forest management activities are vital to timber resources while simultaneously continuing to make use of them. In the same way, fish stocks are also renewable resources. If excessive fish catches are not restrained, fish stocks will be lost but if fish catches are restricted to amounts below what is necessary for them to replenish, then supplies can be sufficiently preserved as they grow back.

Figure 2.2.2 depicts standard growth trends for renewable resources. Consider a given stock of fish in a fishery as an example. In this case, the horizontal axis, which serves to quantify the renewable resource stock, measures the fish population. The vertical axis corresponds to the amount of stock growth and measures changes in the fish population numbers. The growth curve depicts the relationship between the fish populations and population changes at a given time. When the fish population is small, the amount it grows by is also small. On the other hand, provided the fish population is not too large, the growth rate increases as population numbers do. That is to say, too large a fish population renders existing feed stocks insufficient and inhibits the growth rate. Fittingly, the growth curb forms an upside-down "U" shape, as shown in the graph.

In times when no one fishes at all, the growth rate would be positive and the fish population would continue to grow. In said scenario, fish stocks would eventually reach point S_{MAX}, where the growth rate becomes zero and the population numbers level out. Yet when excessive fishing leads the number of fish caught to surpass the number of fish born, the fish population decreases gradually. Conversely, if the growth rate exceeds catch amounts, then fish stocks increase gradually. Finally, if the growth rate and catch amounts are equivalent, then fish stocks decrease via fishing activity by the same amount that they are replenished via propagation, and stock

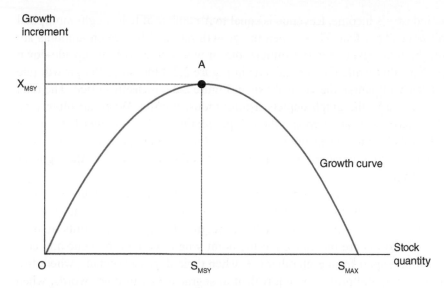

FIGURE 2.2.2 Natural growth increments of renewable resources

amounts can be preserved unchanged through the following year. It is therefore necessary to ensure that the highest possible growth rate (i.e. the rate of fish propagation depicted at point *A*) is preserved in order to succeed at maximizing profits while maintaining a stable fish population. These circumstances call for what is known as the maximum sustainable yield (*MSY*), depicted by X_{MSY}, which determines the number of fish to be caught.

The next aspects to consider are fishing industry benefits. Figure 2.2.3 depicts fishing industry revenue and expenditures. The total revenue curve on the graph indicates

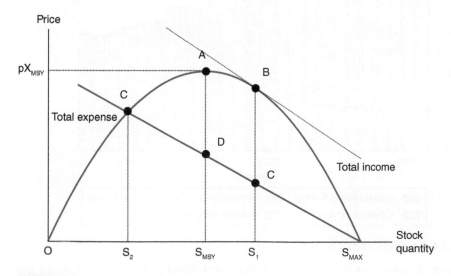

FIGURE 2.2.3 Open access and overuse

the fishing industry's income. Revenue is equal to X number of fish caught multiplied by the unit price P per fish. Thus, when the growth rate and fish catch amounts are equivalent, the total revenue curve mimics the growth curve with an upside-down "U" shape. Note that while the vertical axis in Figure 2.2.2 measures the growth rate (or the fish catch amounts), the vertical axis in Figure 2.2.3 measures price. The total expenditures curve in the graph depicts fishing industry costs. When absolutely no fishing takes place, costs are zero and the fish population is S_{MAX}. Greater fishing and larger catch loads lead to decreasing fish stocks and increasing operational costs over time. As such, the total expenditures curve proceeds downward and to the right, as shown in the graph.

The profits earned by the fishing industry are equal to total revenue less total expenditure. When the maximum sustainable yield is adhered to, total revenue reaches a height of $S_{MSY}A$, total expenditure reaches a height of $S_{MSY}D$, and profits amount to segment AD. Profits are maximized at the point where the total revenue and cost curves are furthest apart from each other (i.e. when their slopes are equal), where fish stocks amount to S_1 and profits are equivalent to segment BC. In other words, when sustainable catch loads are maximized, the fish stocks reach S_{MSY}, and when profits are maximized, stocks drop below S_1.

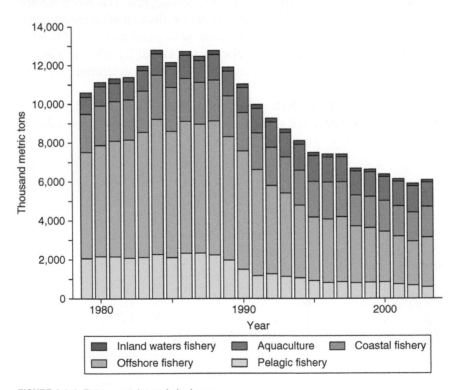

FIGURE 2.2.4 Fishery catch trends in Japan

Source: *Annual Statistics of Fishery and Fish Culture 2003,* Ministry of Agriculture, Forestry and Fisheries of Japan, 2003

Now assume that fisheries are open to free, unrestricted access. Maintaining a fish population of S_1 is essential to maximizing profits. However, even if a given party were capable of limiting its catch loads, its would still miss out on profit-making opportunities should it decide to do so, and moreover, there would be a high chance that other parties would not limit their catch loads and pursue the highest profit accruals possible. The idea of this compels many to persist in continued, unrestrained fishing, causing catch loads to increase until total revenue and total expenditure become equal and profits drop to zero (as depicted by point E). Subsequently, the fish population shrinks to S_2 levels, and fishing organizations experience a deficit regardless of the operational adjustments they may pursue.

In this way, the tragedy of the commons arises when resource use is openly accessible and completely unchecked. Thus, appropriate usage controls are essential to renewable resource preservation. Consider the case of the Japanese fishing industry. The fishing law enacted in Japan restricted the legal authorization of fishing for profit to a limited number of entities. It also put fishing boat restrictions, fishery limitations, fishing time restrictions, fish size restrictions, and other such conventions into place. However, such regulations ultimately act as barriers to entering the fishing industry and do not necessarily help to suppress catch loads. This is apparent from Japan's peak catch loads in 1984 and decreasing loads thereafter.

Then, in 1997, the *Act on the Preservation and Control of Living Marine Resources* established total allowable catch (TAC) loads. TAC promotes resource preservation by dictating the upper limit of catch loads for each type of sea life, ultimately enabling each to procreate efficiently. The act also obliged fishing organizations to publish their catch amounts, and those that surpassed allowable limits were banned from further fishing activities. It was this latter addendum that insured that TAC was not surpassed.

In 2001, the above law was revised to include general regulations on the amount of effort put into catches. To be clear, this total allowable effort, or TAE, is a system that regulates the amount that can be invested into fishing operations and it effectively bolstered the capacity of fisheries to replenish themselves. In scenarios where the aim is to restore fish stocks to a specified level, should one fisherman reduce fishing vessel count while another increases, resource replenishment goals could not possibly be realized. In order to avoid said scenarios, regulations were placed on fishing boat and other catch-related total investments.

Open access to renewable resources leads to their complete depletion in the manner described by the tragedy of the commons. Accordingly, economically efficient regulations are indispensable to appropriate renewable resource management.

LEARNING POINT: INDIVIDUAL TRANSFERRABLE QUOTAS (ITQs)

Aiming to restore fish stocks, Japan implemented a policy which restricted fishing industry total catch loads and investment amounts. However, upon regulating the amount of fish caught, the distribution of catch loads among all fishing entities remained problematic. For example, if the total permissible fishing load was distributed equally across all fishing organizations, this would allot catch loads even to organizations with low productivity, which does not lead to productivity improvements for the industry as a whole. Individual transferrable quotas (ITQs) for addressing this issue have been implemented in various places around the world. ITQ systems allot each fishing entity permissible fishing loads, and excess catch amounts that a given entity possesses can be traded with others. In this system, highly productive organizations can purchase additional load permissions from others in order to increase the amount that they are allowed to catch, while relatively unproductive organizations can sell their loads to others in order to earn profits. Through this ITQ system, economic efficiency was achieved hand in hand with regulated catch loads. In fact, New Zealand has had an ITQ system in place since 1986, and as resource restoration is continuously attainable, the fishing industry has improved its productivity.

SUMMARY

Common pool resources are owned and often managed collectively by all members of a given locale or society. Should they be open to free and unrestricted access, they run the risk of being completely depleted as depicted by the *Tragedy of the Commons*. At the same time, should locales set resource use rules and restrictions, common pool resources could be appropriately utilized and preserved. As most renewable resources are essentially common pool resources, it is imperative to efficiently and effectively manage their access and consumption.

REVIEW PROBLEMS

1. Research an example of the commons. Then, look up whether or not the tragedy of the commons has occurred for it, as well as why it occurred.
2. Though Japanese national parks are free to enter, do you think that they should have entrance fees? Why or why not?
3. The growth function for fish stocks is $x = -S^2 + 10S$. S is the fish resource stock, while x is the resource growth amount in tons. Furthermore, the price for fish resources is ¥1/ton. The total cost (TC) of fishing is $TC = 20 - 2s$. Keeping the above in mind:

    a. Determine the maximum sustainable yield (MSY).
    b. Determine the fish resource stock if open access fishing occurs in this fishery.
    c. Determine the fish resource stock if the fishing industry profits are maximized.

Section 2.3: Public goods and free riding

Public goods and environmental problems

A large portion of the natural environment, from its forests to its atmosphere, have the nature of public goods in the sense that no single entity can monopolize their use. Biological diversity (or *biodiversity*), is a prime example. A great many plants and animal species inhabit tropical rainforests, but because these habitats are rapidly disappearing, many species have gone extinct while others are threatened by the same fate. Considering this, it is safe to assume that there are many people who believe that tropical rainforests ought to be preserved. However, the value of the biodiversity in these tropical rainforests is not accrued by a single or limited number of benefactors, extending to all people across the world. Therefore, it is possible for those who do not financially contribute to such undertakings as preserving ecosystem diversity can still reap the benefits that biodiversity provides. In this way, the natural environment is similar in essence to a public good. Scenarios where some people completely avoid paying their share of the cost of certain (natural) benefits while others shoulder the cost burden themselves depict a behavior commonly known as *free riding*.

In economics, *public goods* are defined as resources that are non-excludable and non-rivalrous in nature. *Non-excludability* is the quality that a good or resource has when its quantity cannot be restricted easily. To gain some perspective into the nature of this concept, consider the case of personal automobiles. Only those who purchase a vehicle with their own currency receive a key for it, thereby making the vehicle free to use for the owner exclusively. In contrast to this, any person or entity can freely emit CO_2 or other exhaust gases into the atmosphere, and it is impossible to restrict such activity to specific entities. The latter case, being a prime example of non-excludability, demonstrates how restricting non-exculdable resource consumption to those who pay for them is practically impossible, and even collecting fees is difficult.

Next, *non-rivalry* is the quality possessed by a good or resource when, regardless of how much people consume it, the amount available for consumption does not decrease. For example, imagine a particular forest is developed into living space and two homes are constructed and marketed. Here, if one person purchases one of the homes, other potential buyers only have the option of purchasing the remaining home. In other words, the homes are rivalrous, so their available quantity decreases as other home buyers acquire them. Compare this example to one in which housing development never gains momentum and the forest is designated a public park. Those who visit the park may be able to enjoy the natural scenery that the park provides, but so long as the number of visitors is within a certain upper limit, spectators can enjoy the scenery even if more people visit. In other words, since scenery is non-rivalrous in nature, even if the number of users increases, consumption levels are unaffected.

Table 2.3.1 separates private goods and public goods along the lines of non-excludability and non-rivalry. *Private goods* are both excludable and rivalrous by nature.

Table 2.3.1 Excludability and rivalry

	Excludable	Non-excludable
Rivalrous	Private goods (automobiles, groceries)	Common-pool resources (fish stocks, timber, coal water)
Non-rivalrous	Club goods (cinemas, private parks, satellite television)	Public goods (ecosystem, climate change policy, natural defense)

Automobiles, groceries, and other common goods are private goods. *Public goods* are both non-excludable and non-rivalrous by nature. Climate change policy and activities to preserve the ecology produce benefits for the entirety of humanity, so they are viewed as public goods. *Common goods* (i.e. *the commons*) are non-excludable and rivalrous by nature. Fish stocks and other types of common resources can be used by anyone if there are no regulations in place. However, if, say, the number of fishermen increases by too much, then the fishing stock, as a resource, depletes. In this way, common-pool resources yield diminishing benefits as the number of users increase and crowding occurs (refer to Chapter 5). Finally, *club goods* are excludable and non-rivalrous by nature. For example, cable television contract subscribers have access to television programmes, and those who are not under a contract can be excluded from access. However, unlike common goods, even if the number of contractors increases marginally, the amount of programmes that all subscribers can enjoy does not diminish.

Private as well as club resources, both being excludable by nature, are conducive to properly functioning market mechanisms mainly because only consumers who have paid a fee are supplied a given good or service. On the other hand, in the case of public and common resources, even those who have not paid for use can partake in said good or service, giving rise to the free-rider phenomenon and market failures.

The optimal supply of public goods

Next, consider the most appropriate supply of public goods, as portrayed by Figure 2.3.1. The focus here will be on climate change policy, and to keep this example simple, assume that there are only two residents who both benefit from climate change policy. To pull off a successful, environmentally friendly policy, costly electricity generation systems such as wind energy must be employed, further adding to policy costs. The marginal cost (*MC*) curve on the graph indicates the necessary cost per each unit decrease of a given greenhouse gas. While climate change policy, as a public good, does indeed continue to increase to the extent that greenhouse gases diminish, it also brings with it rapidly increasing costs. Thus, the marginal cost curve proceeds upward and to the right.

Conversely, the marginal benefit (*MB*) curve on the graph indicates the additional benefits acquired by residents per single unit increase in climate change policy. Although climate change policy impacts all citizens, environmental valuation varies

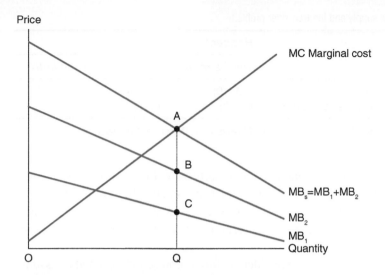

FIGURE 2.3.1 Optimal supply of public goods

from person to person, so the benefit of climate change policy also varies with the person. MB_1 represents the first resident's marginal benefit, while MB_2 depicts that of the second resident. For example, when policy implementation is at levels equivalent to Q on the graph, per marginal increase in policy, the first resident only receives benefits amounting to CQ, while the second resident gains amounts worth BQ. Since there are only two residents, the total social benefit is the sum of the benefits earned by each resident. In other words, $CQ + BQ$ = total social benefit AQ. By combining the marginal benefits of both residents, it is possible to graph MB_S, the marginal social benefit curve.

Socially optimal climate change policy levels are realized where the social marginal benefit and marginal cost are equivalent. When marginal benefits surpass marginal costs, then the net benefits of proceeding with further policy measures increase. Conversely, when marginal benefits fall below marginal costs, the only way to increase net benefits is to curtail the overly expensive policy measures. Point A, where the social marginal benefit and the marginal costs are equivalent, includes socially optimal conditions where the policy, as a public good, is provided at optimal level Q.

Public goods and the free-rider problem

However, it is not simple to achieve optimal levels as described above. To do this, marginal benefits of a given public good gained by individual residents must be calculated. Individuals must therefore be asked how much he or she is willing to pay per marginal increase of climate change policy. Then, a supply cost must be appropriated to the public good in question to cover the monetary worth of the benefits that each resident reaps. Yet, if the cost imposed on a resident is to be determined by how they

Table 2.3.2 Public goods supply and the free-rider problem

		Resident 2	
		Tell the truth	Under-express the truth
Resident 1	Tell the truth	(10, 10)	(0, 20)
	Under-express the truth	(20, 0)	(5, 5)

Note: The values in the parentheses are (Benefits of Resident 1, Benefit of Resident 2)

respond to questions about the benefit they actually gain from the climate change policy, they would be incentivized to falsely depict the benefits they anticipate to gain as being worth less than their actual value. For, even if one did not answer truthfully, others will bear the financial burden, so the respondent is thereby capable of free riding on climate change policy.

Table 2.3.2 depicts this kind of free-rider problem among public goods. Assume two local residents are to be asked about the benefits of climate change policy. Each person can choose to either honestly express or undervalue the benefits they receive. However, if either one of them undervalue-expresses, then policy measurements can only be implemented to the lowest standards possible. For example, numerical values (20, 0) represent the benefits to Resident 1 in the left column and Resident 2 in the right column. In this scenario, one resident gains benefits worth 20 points, while their counterpart receives zero benefits.

If both people truthfully answer, the numerical values become (10, 10) as both gain benefits worth 10 points. However, if only the first resident understates the value, then the points are distributed as (20, 0). This means that the first resident, receiving high benefits while minimizing their own cost burden, relies on their counterpart to bear the brunt of the expenses. The first resident doesn't properly express the value and the climate change policy is not sufficient, and the second resident's benefit drops. On the other hand, assume that the second resident understates the value and the points are distributed at (0, 20). While they could both understate the value and suppress their individual cost burdens to the lowest feasible levels, the climate change policies can also only be implemented at the lowest levels, and both residents gain benefits worth (5, 5).

As this point, consider the stance of the first resident. If the second resident answers truthfully, and the first resident does as well, then they both receive benefits amounting to 10. Yet, if Resident 2 answers truthfully and Resident 1 lies and undervalues the worth of the policy, he or she can earn benefits equaling 20. Conversely, if Resident 2 decides to lie about the value, and Resident 1 tells the truth, then Resident 1 gains zero benefit. And, if Resident 1 lies along with Resident 2, then they both only gain benefits worth 5. Accordingly, Resident 1 gains the most benefit from lying. And, Resident 2 thinks the same thing, which results in them both lying. In this way, there are incentives to free riding, and it is difficult to realize optimal levels of public goods.

SUMMARY

As public goods in nature, ecological preservation and climate change policies are non-excludable and non-rivalrous. Since public goods incentivize free riding, market mechanisms do not function efficiently. Due to this, when environmental policies involve public goods, not only voluntary measures among consumers and businesses, but also government-mandated controls such as environmental regulations and taxes must be imposed.

REVIEW PROBLEMS

1. Think of an environmental problem that involves public goods, and investigate how the free-rider problem may arise.
2. About 30% of the forests in Japan are managed by the government as national forests. Do you think they should be privatized? Why or why not?
3. Consider the limitations of self-imposed climate change countermeasures among businesses.
4. Perform the public good supply experiment outlined in the **Learning point**. Then, consider what is necessary to avoid the free-rider problem in the experiment.

LEARNING POINT: AN ECONOMIC EXPERIMENT WITH PUBLIC GOODS

The following simplified economic experiment highlights the free-riding issue as it relates to public goods. This experiment must be performed in a group. Figure 2.3.2 depicts the sequence of events for the game. First, make a group of about three to six people. In this example, three people participate.

The experiment is comprised of ten stages. At the start of each stage, every player is handed 15 points. Then, each player decides how many of those 15 points they will apply to payments for public goods. In this example, Player A contributes 15 points, Player B provides a mere 5 points, and player C gives nothing at all. The government then collects the points uses them to invest in aims related to public goods, in this case climate change policy. As a result, the government hands out an amount equal to half of the collected points to players in the form of benefits accrued from climate change policy. In this example, because 20 (= 15 + 5 + 0) points were collected, half of this, or 10 points, will be distributed to each player. This signifies an equal amount of benefits from climate change policy and other public goods being distributed among everyone. The number of points that each player possesses, comprised of

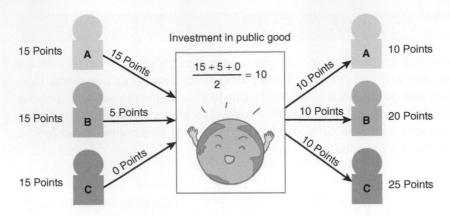

Experimental procedure

1. Create a group of 3 to 6 members.
2. Provide each player with 15 points.
3. Allow each player to decide how many of their 15 points they will contribute to the public good.
4. Invest the points collected from each player into the public good and provide each player with half of the total number of points collected.
5. Repeat steps 2 to 4 10 times.

FIGURE 2.3.2 Experiment with supply of public goods

the number of points they previously had in their hand plus what was received from the government policy, becomes the player's profit for the stage. In this example, Player A initially paid the total 15 points they possessed, leaving them with 0 points at first, but then acquiring 10 points from the government policy, resulting in a mere 10 point profit. Player B, having only paid 5 points initially, still possessed 10 points on hand and then gained an additional 10 points from the government policy. Thus, they acquired 20 points overall. Player C, possessing 15 points after not having paid anything initially, accrues a total of 25 points after acquiring 10 points from public goods via government policy. Overall, Player C is a basic example of how the free-rider problem originates when people gain benefits from public goods (e.g. through policy effects) while not paying them.

At this point, consider what would happen if all players decide to cooperatively invest all of their points into the environment. Thus, as each player puts forth 15 points initially, the total points collected for the public goods are 45, and half that amount, or 22.5 points, is handed out to each player. Since everyone paid all of their initial points, they have none left, but each player's final profits are 22.5 points that they gain as benefits from the public good. This means that if everyone cooperates, everyone can increase their profits from their original 15 to 22.5 points. In other words, it is possible to make progress with climate change policy should everyone endeavor cooperatively, leading to profit increases for everyone across the board.

Conversely, consider what would happen if all players completely reject paying anything for public goods. The three players hold on to their initial 15 points, and since there is no investment in public goods, no public benefits arise. The result is that the initial 15 points held by each player becomes their final profit. In other words, if everyone free rides, the climate change policy is unsuccessful, and it is impossible to raise gains.

If only one person doesn't pay for the public goods, and the other two people pay the full price, what would occur? One person is left with 15 points in their possession, while the other two, having paid all of their points, are left with 0 points. Since the two who paid collected 30 points, everyone receives 15 points in benefits. In the end, the person who initially kept their 15 points and received the 15 points in collective benefits ends up with 30 points. On the other hand, the other two people, being left with 0 points, only acquire 15 points from public good benefits in the end. In short, even when other people worked together for the cause of the public goods, if an individual undermines that by not cooperating, they end up with greater final profits than those who actually made efforts.

Therefore, while there do exist incentives to earn high profits through free riding with public goods, what would result if things actually occurred as in this experiment? Figure 2.3.3 depicts such a feat. The horizontal axis corresponds to the stage, while the vertical axis corresponds to the percentage of the initial 15 points that are invested in public goods. The outcome is largely influenced by the number of players and the amount of benefits that can be received, and as shown in the figure, they generally form a line that proceeds downward and to the right. Through cooperation in early stages, high benefits from public goods can be anticipated, so people invest a comparatively large number of their points into public goods. However, as time goes on, players realize that others are free riding, and even though they themselves are investing in public goods, they are still losing out overall. This leads individuals

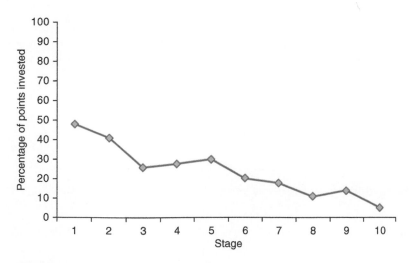

FIGURE 2.3.3 Example of experiment results

to decrease their investment amounts. Consequently, as portrayed in the graph, the number of free riders increases as the stages progress.

As we can see from this experiment, because there are incentives to free ride with public goods, voluntary actions among individuals alone are not sufficient in dealing with the cost burden of public goods, rendering it difficult to supply appropriate amounts. Therefore, as is the case with ecosystem preservation and climate change policy, when environmental policies involve public goods, the voluntary consumer and corporate action must be applied alongside governmental environmental regulations, environmental taxes, and a host of other conventions.

CHAPTER 3

Fundamental theories of environmental policy

Chapter overview

This chapter lays out the steps to achieving environmental issue resolutions. Ideal environmental policies can be classified into two general groups: the command and control type of direct regulations that have been traditionally employed in policy processes, and the economic measures that have recently come to prominence. One example of a direct regulation is explicitly defining permissible sulfur dioxide emission levels and prohibiting excessive emissions. As one can see from this example, direct regulations are straightforward and authoritative as via environmental management mandates. Economic methods, on the other hand, indirectly bring about environmental oversight by appealing to the rational decision-making processes of each relevant economic entity. Accordingly, this chapter will introduce the special features of both methods.

Chapter content

Section 3.1—Many countries, including Japan, employ environmental policies centered around direct regulations. This section sheds light on the impacts that direct regulations have and also compares them with the market mechanisms utilized in economics-based methods.

Section 3.2—This section introduces the procedures inherent in environmental tax and subsidy policies. Subsequently, scenarios when tax burdens are borne by consumers vs when they are borne by producers are discussed. Finally, considerations of the various short-term and long-term effects that both environmental taxes and subsidies have are provided.

Section 3.3—This section introduces the Coase theorem as the principle that underlies problem solving via direct negotiations. The conditions necessary for related parties to resolve externalities are outline. Furthermore, debates regarding environmental rights are introduced.

Section 3.4—Under emissions trading systems, organizations are able to exchange the rights to emit specific amounts of specific pollutants. This section explains types of problems that arise with the implementation of emissions trading systems and introduces examples from across the globe.

Section 3.1: Direct regulations and market mechanisms

Direct regulations and economic measures

As previously mentioned, when externalities exist, various policies are needed to keep environmental pollution at appropriate levels. Below are some of the premier environmental policy examples:

1. Direct regulations of pollutant emission levels (e.g. command and control or *CAC*),
2. Taxes imposed on pollution emissions and subsidies provided for pollution emission reductions,
3. Deals made via direct negotiations,
4. Improvements via tradable emission rights markets.

The emissions cap and trade, environmental tax, and environmental subsidies systems that employ *market mechanisms* are known as *economic methods*, and as such, they emphasize incentives for economic entities. In recent years, as the limitations and inefficiencies of direct regulations have become increasingly clear, the advantages of economic methods are often preferred. Outside of the previously mentioned methods, solutions through direct negotiations among parties could also be grouped into economic-based methods.

What are direct regulations?

Traditionally, command and control-style direct regulations have been proposed to resolve environmental problems. By imposing such direct regulations as operational restrictions and mandates stipulating acceptable behavior on polluters, governments aim to regulate pollution emissions. Two examples of these are *total emission amount regulations* and *emission criteria regulations*. Total emission amount regulations are systems that fix the upper limits of permissible pollution emission levels.

Emission criteria regulations are systems that categorically restrict contamination percentages among production emissions. More specifically, they set input coefficient restrictions of environmental resources (indicators of environmental resource utilization that could also be called the *pollution emission coefficient*) per unit of production.

Japan, along with many other countries, employs direct regulations because they make policy targets generally easy to understand. Take dioxins for example. Incinerators are the primary dioxin emission sources, and since they cause damage to vegetation and other aspects of the surrounding environment, they bring about social issues that could potentially have serious impacts on the health and livelihoods of people.

Considering the above, the *Act on Special Measures concerning Countermeasures against Dioxins* was enacted in 1999 to establish air, water, and land contamination standards. Since the negative impacts that dioxins have on the environment were well known, regulations for prohibiting dioxin emissions easily gained broad-based support. However, many of the contaminants that negatively impact the environment are the byproducts of manufacturing processes of profitable goods. Thus, it is crucial to note that enforcing emissions prohibitions could simultaneously mean the prohibiting or restricting the production of important goods. Accordingly, the following sections will introduce the methods that lead to appropriate restriction levels.

Optimal pollution levels

Ideal management of the environment brings about pollution emission amounts that are at socially optimal levels. Optimal levels exist when the difference between total pollution damages deducted from economic gains from polluters is maximized. Yet, how should socially optimal emission levels be determined in a specific and justifiable manner? First, society must collectively decide the extent to which pollution should be curbed. Then, which entities will be obliged to cut emissions, and to what extent, must be determined. As one may expect, these distinctions influence policy choices. For now, consider the first of these operations.

Figure 3.1.1 depicts the marginal revenue (*MR*) and marginal cost (*MC*) functions of a given business' manufacturing processes. Marginal revenue is the additional revenue that can be earned per each unit increase in production. Assuming this example takes place in a perfectly competitive market, the marginal revenue is equal to the market price P'. This is because within perfectly competitive markets exist a great number of both sellers and buyers, and the price of a given good is defined by the equilibrium point of the entire market, where supply and demand are equal. Therefore, the marginal revenue of each business is the revenue received when one additional unit of production is sold, which is the same as the market price of the sold unit.

As explained in Chapter 1, Section 1.3, the marginal cost is the additional cost generated per single unit increase in production. Furthermore, the marginal cost curve correlates with the supply curve for the products that a business provides to the market.

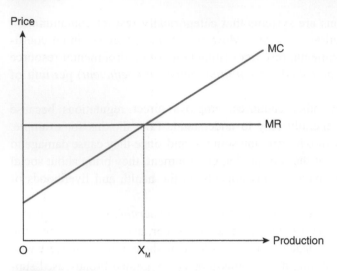

FIGURE 3.1.1 Business revenues and costs

The difference between price and marginal cost, or the marginal benefit (*MB*), is the additional benefit gained per single unit increase in the production and sale of a product. Figure 3.1.1 shows the vertical distance from the graph where the marginal revenue and marginal cost curves intersect. Placing the marginal benefit of this point on the axis, Figure 3.1.2 shows the marginal benefit curve.

Now, as pollution is yielded through the production of said product, suppose the amount of contaminants and waste increases in proportion to production increases.

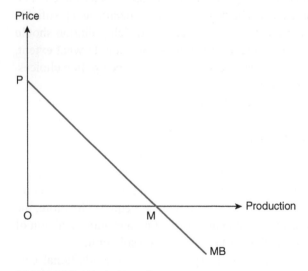

FIGURE 3.1.2 Marginal benefit curve

In other words, while this unit will consistently measure production levels along the horizontal axis, the horizontal axis will also depict waste levels.

Take another look at Figure 3.1.2. When production levels equivalent to X_M are exceeded, the marginal cost surpasses the price, leading to a negative marginal benefit, and continued production at levels above X_M yields losses. Therefore, optimal production levels are realized at X_M, where company profits are maximized. At this point, businesses are not, on any level, considering the amount of waste being produced. And, since company profits are equivalent to the marginal benefits from point O to X_M, they amount to the area of $\Delta OP'X_M$.

Next, imagine a scenario in which only two companies, Company 1 and Company 2, emit pollution through production. Figure 3.1.3 depicts the marginal benefit of each company. Additionally, the shape of each company's marginal cost (and marginal benefit) curve differs based on technological reasons.

Here, if both companies were not regulated, how much would they pollute? When there are no regulations on waste production, production increases lead only to additional profits, and each company will continue to increase production until marginal benefits become zero and profits are maximized. But, as a result of production, pollution emissions contaminate the environment.

In the case of Figure 3.1.3, both companies maximize profits by producing 100 units (for example, 100 tons). Assuming no other companies exist, total production levels of society are Company 1's 100 + Company 2's 100 = 200 altogether. Combining the marginal benefit curves of both companies ($MB_1 + MB_2$) amounts to the total marginal benefit to society, depicted by horizontally progressing line in Figure 3.1.4.

From here, we make assumptions about the proportion of waste that is produced alongside production by viewing the horizontal axis in Figure 3.1.4 as an indication of pollution levels, allowing us to consider the marginal externality costs that arise from pollution. Since the damages inflicted become more severe as pollution levels

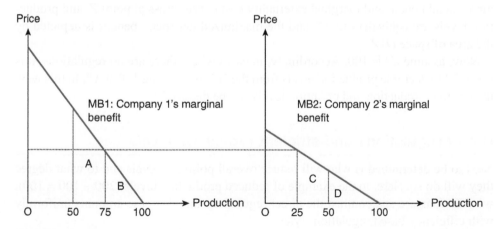

FIGURE 3.1.3 Effects of direct regulations

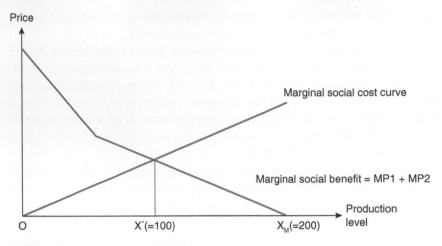

FIGURE 3.1.4 Socially optimal production levels

increase with production, one could assume that the marginal externality cost of environmental pollution increases to the extent that production increases. In other words, the marginal externality cost curve proceeds upwards and to the right.

As previously explained with Figure 3.1.2, the area beneath the marginal benefit curve depicts company profits, so in this current example, it represents the total profits of both companies. Additionally, the area beneath the marginal externality cost curve depicts the externality costs. To determine the total social net benefits, one must subtract the externality costs from the total profits of both companies. This amount is the maximum production level, but for society, it is also the optimal production level. In other words, socially optimal production levels are derived from the maximization of utility, where the remainder between the net detriment of pollution emissions is deducted from the boons of economic activity. In Figure 3.1.4, this is where the marginal social benefit and marginal externality cost curves cross at point Z, and production levels are equivalent to X^*, and the maximized net social benefit is depicted by the area of space OYZ.

Now, assume X^* is 100. Accordingly, in cases where there are no regulations, it is essential to decrease production levels from the 200 at X_M to the 100 at X^*. In this way, the optimal production and pollution levels can be deduced.

Direct regulations and efficient regulation levels

Next to be determined is who will reduce overall pollution levels and to what degree they will do so. Here, in the example of reduced production levels (200 − 100 = 100), we will compare and contrast the direct regulations that decrease amounts uniformly with efficiency-based regulation styles.

The first focus will be on direct regulations. Consider, for instance, that both companies uniformly cut production levels to 50 from 100 in order to reduce pollution. While company profits do indeed decrease alongside decreases in production, since there is a gap in the level of technology among the companies, the amount by which profits decrease for each company is different. The reduced profits for the two companies can be determined by measuring the area underneath the marginal benefit curve for each company in Figure 3.1.3. In this case, when production dropped from 100 to 50, Company 1 experienced a decrease equal to area $(A + B)$, while Company 2 witnessed losses equivalent to the area of D. In other words, the total social profit losses amounted to $(A + B + D)$.

At this point, assume that reduction amounts can be set along the lines of each company's technology level and not via uniform, overall reduction regulations. Consider, as depicted in Figure 3.1.3, Company 1 possesses more advanced technology, and profits largely decrease in the event of production decreases. Assume that conditions for Company 1, experience a large profit decrease from 50 to 25, placing its production levels at 75. In this case, the profits decrease by as much as the area of B. Now, for Company 2, which does not lose many profits in this scenario, makes reductions of not 50, but 75, leaving production levels at 25. Thus, the profit drops to $(C + D)$. The total social reduction amounts, similar to the direct regulation scenario, are equal to 100, and the profit losses equal to $(B + C + D)$ amount to the cumulative losses of both companies. If the losses in $(B + C + D)$ are less than the losses in $(A + B + D)$ found in the direct regulation scenario, then this method is said to be efficient. By comparing the two, one can see that

$$(A + B + D) - (B + C + D) = A - C > 0$$

so it is clear that $(A + B + D) > (B + C + D)$. In other words, one advantage of direct regulations is that to the extent that a company preserves a portion of its production, the initial schedule for clearly executing the government's initial goals can be realized. However, beyond uniform emissions reductions, invoking reduction methods according to technology levels is said to be more efficient.

Now, it is also possible for governments to reach optimal reduction levels in accordance with the technology levels among businesses without directly setting the goals. For instance, assume that per single unit of production, an environmental tax equal to t^*, found in Figure 3.1.4, is imposed. Here, profits for Company 1 are maximized when production levels reach 75. The reason for this is that if production levels were raised past this point, environmental tax payments would surpass marginal benefits, and profits would decrease. Similarly, profits for Company 2 are maximized when production levels reach 25. If reduction levels amount to 25 + 75, then the socially optimal production levels can be realized.

By deciding the appropriate tax rate in order to guide production to appropriate levels, the socially optimal production levels can be implemented. While the following section will explain the measures for determining environmental taxes in detail, at this point, be sure to note that appropriate tax rates eliminate the fear that profits are being excessively lost by society.

Voluntary controls

Voluntary controls (also known as *voluntary approaches*), through which the business world imposes regulations naturally and without direct oversight from the government, are additional regulatory options. These could exist in the form of pollution reduction environmental policies in which goals are determined, executed, and enforced by the emitters themselves. They are relied upon to overcome challenges related to determining and achieving goals that arise when government standards do not suit business needs or capabilities. Since they are voluntary and industry-created, they generally reflect overall feasibility and indirectly invite government participation.

Consider one representative method. Since it is essential to institute production levels conducive to significant pollution reductions, the business world considers to what degree voluntary restrictions are effective. Then, the goals are decided along the lines of the goals set by actual emitters themselves. However, since the aims of these goals are often weakly linked to policy, these policies are often thought to be unreliable for actually accomplishing said goals. Therefore, there can be enormous obstacles to overcome when relying on voluntary regulations derived from business leadership objectives.

Moreover, there are some cases when governments play extensive roles in voluntary regulations, strict environmental goals are set, and emission reductions are achieved. These cases are considered to be more efficient than standard environmental policies. Consider the Top Runner Program of Japan that is credited for successfully spurring on emissions reductions.

Following the oil shocks of the 1970s, and especially recently with global environmental problems in the background, there have grown to be great expectations for energy conservation, and energy efficiency improvements among electrical appliances are now in high demand. With these conditions as incentives, the *Top Runner Program*, 1998 was introduced as a means for setting energy consumption efficiency standards among automobiles, household electrical appliances, and other mechanical devices in order to promote energy conservation among civilian and transportation sectors. The Top Runner Program, as a basis for evaluating the relative energy utilization capabilities of all products on the market, became an assessment criterion for further improvements in efficiency.

For mechanical manufacturers, technological and financial burdens arise in the absence of new technologies. Furthermore, such products are valued more than conventional ones. Therefore, in the Top Runner scenario, there were many incentives for vendors, consumers, and manufactuers to develop technology. First, stores that go out of their way to share information and promote the sale of such products were publicly acknowledged, and initiatives for venders of energy related products were promoted. Furthermore, it utilized a labeling system to indicate the energy conservation capabilities of electrical appliances, which spurred on the diffusion of highly efficient devices. Additionally, when manufacturers failed to reach their goals, the Ministry of the Economy provided advice, and if the company failed to

Table 3.1.1 The Top Runner Program

Equipment name	Energy Consumption efficiency percentage improvement	Energy Consumption efficiency percentage improvement (as initially anticipated)
Television receiver	25.7% from 1997 to 2004	16.4%
Videotape recorder	73.6% from 1997 to 2003	58.7%
Air conditioner	67.8% from 1997 to 2004	66.1%
Electric refrigerator	55.2% from 1998 to 2004	30.5%
Electric freezer	29.6% from 1998 to 2004	22.9%
Gasoline passenger automobile	22.0% from 1995 to 2004	23.0% from 1995 to 2010

comply, it was publicly announced, and they were given orders conducive to reaching their goals.

To date, the Top Runner Program has gone through one round, targeting television reception, video tape recorders, air conditioners, electric refrigerators, and freezers. As indicated by Table 3.1.1, efficiency improvements surpassed what was originally anticipated. In other words, the regulations that were implemented had largely encouraged technological progress.

However, while it is said to be effective for spurring on energy conservation, whether it was an efficient system despite the exorbitant business and government guidance costs that it brought about, how efficient it is compared to the energy taxes and other market mechanisms (to be explained in subsequent sections) is still up for debate.

SUMMARY

Direct regulations are those by which the government has a direct hand in the behavior of pollution emitters or commands that such actors perform in a particular manner. One method for achieving this is to set the upper limits of pollution emissions. Direct regulations make it difficult to implement socially optimal emissions levels, but it is possible to utilize economic means to achieve efficient environmental management.

REVIEW PROBLEMS

1. Explain how to determine optimal pollution levels.
2. Describe why direct regulations make it difficult to achieve optimal emission levels.
3. Explain the processes and results of two different voluntary restraint regulations.

LEARNING POINT: ARE DIRECT REGULATIONS CAPABLE OF EFFICIENTLY REDUCING POLLUTION?

Through direct regulations that fail to make use of economic measures, is it possible to achieve socially optimal pollution levels? In other words, is it possible to allocate production to levels that make marginal abatement costs for each and every business?

First, one must be aware of the marginal emissions cost at pollution sources in order to implement equivalent marginal costs among businesses. To do this, it is essential to understand properly the manufacturing structures of all businesses. However, in reality, as governments are not the businesses themselves and have imperfect information regarding the marginal cost curve of industries' pollution, this is very difficult to accomplish. Consider this from an industry perspective. For an industry, by telling information to the government, if it leads to the industry having rules placed on it, then there is no reason for it to directly provide accurate information. In reality, by reporting to the government of its own accord, businesses manipulate information strategically. For example, depending upon a firm's past emission levels, if the government plans to impose a number of restrictions, it is possible that said firm would intentionally produce reports that purport emissions levels that were much higher than reality. In this way, there are differences in the verifiability and amount of information between that which the government uses to impose regulations and that held by reporting firms that will have rules imposed on them. This is known as *asymmetrical information.*

Especially in cases where unproductive firms are allowed high production levels and where highly productive firms are allowed low production levels, the overall market production costs increase, and it becomes impossible to maximize social profits. In the end, the government must use discretion in setting pollution levels, and as such, it is likely that the government takes into account the influence on each industry and the negotiation ability of firms. What results is that the allotments for large firms with large-scale negotiation capabilities are excessively high, which preserves one part of the economy as a whole, but the social profits decrease. In the end, there are also instances where the government must give legal instructions to firms. Then, there lurks the danger of the firm being unable to strictly follow the guidance and becoming unable to live up to regulations. Furthermore, legal guidance may require excessive costs.

Section 3.2: Environmental taxes and subsidies

Environmental tax

An *environmental tax* is a financial burden, imposed on material emissions that negatively affect the environment, which aims to control pollution emissions. This is not a specific regulation method that sets standards through the mainstream pollution treatment and emissions laws that have been utilized thus far. Instead, it is a general term for the taxes that are implemented to resolve environmental issues through economic measures.

Environmental taxes are implemented according to the pollution content of emissions. Although emitters can yield as much as they choose as there are no upper limits to pollution amounts, their emission levels are measured, and the greater the emissions, the greater the taxes imposed. At the root of these kinds of taxes is an incentive for polluters to reduce their emission levels by as much as possible. In conditions where there is no environmental policy, environmental resources can be used freely, but one is induced to economize when environmental taxes are present. Along with decreasing costs related to materials and labor, in the face of taxes, one would aim to decrease emissions and the costs they bring about as well. In other words, environmental taxes are expected to internalize negative externalities. Moreover, as environmental taxes were purported to make the market take in negative externalities, they have come to be known as Pigovian taxes.

When environmental taxes are implemented, the means of decreasing emission levels are entrusted to the polluter. This is the special feature of environmental taxes, as polluters put forth the necessary effort to make reductions. Emission treatment methods, alterations among production processes, recycling, and switching to comparatively cleaner materials are all pertinent examples. From here, we will explain the environmental tax policies to emphasize the importance of the market. Moreover, in this section, we will explain systems related to energy, and in Chapter 4, Section 4.2, we will outline the implementation of fees for waste disposal.

Optimal environmental tax

Use Figure 3.2.1 to consider changes to production levels when environmental taxes imposed on pollution emitting enterprises are equal to t^*. If taxes are not imposed, companies will produce at levels equal to X_M, where profits are maximized, and pollution emission amounts are determined in accordance with the production levels (refer to the previous unit). At this point, the company earns profits equal to $A + B + C + D$, but pollution that arises alongside production yields externality costs equal to area $C + D + E$ beneath the marginal externality cost curve. The social net benefit thus becomes $(A + B + C + D) - (C + D + E) = (A + B - E)$.

Next, assume that $¥t^*$ is imposed per single unit of production. Here, production levels and pollution emission levels are considered to be proportional, so one could also think of this as $¥t^*$ imposed per single unit of pollution. Under these circumstances, when a company increases production by one unit, it must bear an additional financial burden of $¥t^*$. Accordingly, the business will continue to reduce production

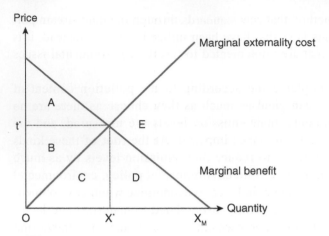

FIGURE 3.2.1 Optimal environmental (Pigovian) tax

levels until the point where the marginal externality cost curve and the marginal benefit curve are equivalent. The reason for this is that if the company were to produce above X^* levels, the economic tax t^* becomes greater for every additional unit above the marginal benefit, making the company profits shrink. Because of this, the company has no incentive to produce above X^* levels. Further still, if the company produces below X^* levels, the marginal benefit becomes greater for every additional unit above the marginal tax, and the company profits increase. Because of this, the company has an incentive to increase profits to X^* levels. Finally, as the environmental tax is imposed, production levels shrink from X_M to X^*. Since production amounts and pollution emission levels are proportional, the tax imposition leads to pollution reductions as well.

Before the tax is imposed, the company had been producing at X_M levels, and the area $A + B + C + D$ beneath the marginal benefit curve equated to profits. After the tax imposition, since the production levels shifted from X_M to X^*, profits worth the area of D were lost. Moreover, as taxes paid to government equated to $B + C$ ($=$ tax amount t^* x production amount X^*), the company was left with profits equivalent to area A only. The amount of tax collected by the government can be allocated to those impacted by environmental problems, and there are also ways not related to environmental problems that they can be used, so net social benefits can be raised. Externality costs become equivalent to area C beneath the marginal externality cost curve. Accordingly, the equation for the social net benefit becomes $A + (B + C) - C = A + B$, meaning that the net social benefit is raised by E compared to the pre-environmental tax $(A + B - E)$ levels.

Improvements in economic welfare through environmental taxes

As we witnessed in the previous section, optimal environmental tax levels can be derived at the point where the marginal benefit and marginal externality cost curves intersect. Recall that here, the marginal benefit is the difference between the marginal revenue and the marginal cost (refer to the previous section). In other words, production levels are optimal when *Marginal Revenue – Marginal Cost = Marginal Externality Cost*. By rewriting this equation, it becomes *Marginal Revenue = Marginal Cost + Marginal Externality Cost = Marginal Social Cost*. Still, in a perfectly competitive market, the marginal revenue is equivalent to the market price, which means that when production levels are optimized, the equation can also be written as *Market Price = Marginal Social Cost*. Because the market price is determined along the demand curve, economic welfare is maximized at the intersection of the demand and marginal social cost curves.

However, companies adhere to the equation *Marginal Revenue = Market Price = Marginal Cost* when choosing production levels, so socially optimal production levels are not realized. At this point, set the environmental tax to levels equivalent to the marginal externality cost when production levels are optimized. By doing this, companies choose production levels in line with the equation *Marginal Revenue = Market Price – Environmental Tax = Marginal Cost*. Here, production levels are optimized, and the market price and the marginal social cost coincide. Furthermore, the pricing adopted by the company deducts the environmental tax from the market price. T (this kind of environmental tax is a Pigovian tax.

It is now necessary to confirm these concepts using the demand and supply curves depicted in Figure 3.2.2. Before the environmental tax is imposed, transactions are made at point E_0, where there is balance as the supply and demand curves cross. At this point, just as we saw in Chapter 2, Section 2.1, the costs yielded are equivalent to the area of $\triangle BE_0F$, and the total economic welfare becomes the area of $\triangle ABE_1 - \triangle E_0E_1F$.

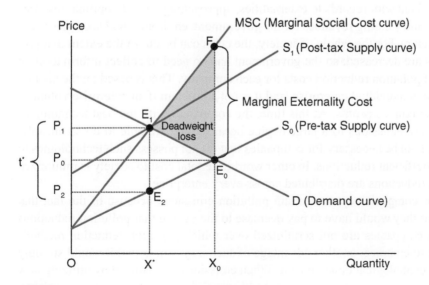

FIGURE 3.2.2 Improvements in economic welfare via environmental taxes

Here, assume that environmental tax t^* is imposed upon E_1E_2 per additional unit of production. Before the tax, when the company increased production by one unit, only the marginal cost went up. However, after the tax, as the single-unit production level increases, the company must bear the environmental tax burden t^* in addition to the marginal cost. Because of this, the company shifts to supply curve S_1, and goods are dealt at point of balance E_1, where the demand curve D and new supply curve S_1 intersect (note the fact that the supply curve depicts the marginal cost, as explained in Chapter 2, Section 2.1).

Next, have a glance at economic welfare after the tax imposition. As the market price rose from P_0 to P_1, the amount that consumers purchase decreased from X_0 to X^*. Here, the consumer surplus is equivalent to the area of $\triangle AE_1P_1$.

At the same time, the price adopted by the producers becomes P_2, the remainder after environmental tax t^* is subtracted from the original market price P_1. At this point the revenue earned by the producers becomes $P_2 \times X^* =$ the area of $\square OX^*E_2P_2$, and when the area of production costs equivalent to the area of OX^*E_2B are subtracted, the producer surplus becomes equivalent to the area of $\triangle E_2BP_2$.

The tax revenue earned by the government becomes $t^* \times X^*$, or the area of square E1E2P2P1. Furthermore, the externality cost becomes $\triangle E_1E_2B$. Accordingly, the post-tax economic welfare can be derived from the equation $\triangle AEP_1 + \triangle E_2BP_2 + \square E_1E_2P_2P_1 - \triangle E_1E_2B = \triangle ABE_1$, and the welfare equivalent to the prior deadweight cost is recovered through the environmental tax.

The merits of environmental taxes

The fact that environmental taxes can bring about efficient pollution reductions in the manner described above is what makes them excellent. Environmental taxes imply the ability to attach a price directly to the right to pollute. By correcting the incentives of those involved with regards to externalities, approaching socially optimal resource distribution, and yielding revenues for the government, environmental taxes raise economic efficiency. For the whole of society, the costs can be cut to the extent that pollution levels are decreased, so the government has no need to collect information on the marginal pollution reduction costs for each company. This is based on the tax rate applied to each individual enterprise, and it is not a problem if an enterprise voluntarily aims to maximize profits. At this time, the companies with the best technologies would decrease emissions to a large degree because they are able to do so cheaply. Yet, it would not be necessary for companies that do not possess such technologies to force such inefficient reductions. In other words, the total cost to society is minimized as pollution reductions are distributed across every enterprise.

Moreover, companies opt to restrain pollution emissions because of the fact that the taxes that they would have to pay decrease to the degree that pollution reductions do. And, as companies are not scrutinized over which pollution reduction methods they choose to embrace, another advantage of such a system is that it would strongly incentivize technological development. When emissions are reduced by utilizing new technologies, double cost reductions along the lines of technology expenses and tax payments act as stronger incentivizing factors than mere adherence to regulations. With this in mind, because environmental taxes lead to tax exemption through considerable endeavors, they are said to incentivize technological development. This includes polluters' aims for developing and improving upon production methods that emit smaller amounts of pollution, developing and improving upon pollution decontamination technologies, and using more environmentally friendly resources.

Are taxes burdens to producers, or to consumers?

According to the extent by which demand is affected by price changes, we can measure the resultant pollution reductions and the size of the burden placed on polluters to achieve such reductions. The numerical extent to which taxes bring about pollution reduction can be found in reflections of the changes. The qualitative effect that taxes have on emission reductions depends on the extent to which polluters react to price changes.

Here, to measure the extent by which supply and demand react to price changes, we consider the price to be elastic. Price elasticity of demand depicts by what percentage demand levels change for every 1% change in price. When the demand level largely changes, it is said to be highly elastic, and vice versa. In other words, if the slope of the demand curve is steep, it is said to be slightly elastic, and vice versa. Figure 3.2.3 (1) depicts a good with slightly elastic demand, while Figure 3.2.3 (2) depicts the opposite trend of a good with largely elastic demand. In both scenarios, the supply curve takes the same form.

First, consider the environmental tax as a burden to the consumer. The consumer suffers a burden equivalent to the cost of $(P_1 - P_0)$ as the price increases from P_0 to P_1. Next, consider the burden to the producer. At a glance, when the market price rises from P_0 to P_1, one might think that the producer does not suffer a price burden. However, since there is an environmental tax imposed, the price that the producer incurs is equivalent to $P_2 = (P_1 - t^*)$. In other words, the producer suffers a price burden equivalent to $(P_0 - P_2)$.

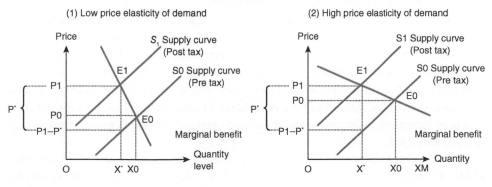

FIGURE 3.2.3 Environmental tax burdens

Compare Figure 3.2.3 (1) and (2). The consumer burden in (2) is less than the $(P_1 - P_0)$ burden suffered in (1). Thus, we can see that the greater the demand elasticity, the less the burden on the consumer.

Gasoline

At this point, in order to reinforce *understanding* of the relationship between the environment and elasticity of demand, we will consider gasoline as a resource. Since areas that do not have trains, buses, and other methods of public transportation readily available cannot replace car use in the event that gasoline prices go up, in this case, we can consider the price of gasoline demand to be relatively inelastic. Accordingly, in areas that have considerable public transportation methods, more people will use such methods in the event that gas prices rise, so we assume in this case that the price of gasoline is quite elastic.

Figure 3.2.3 (1) depicts the demand curve for elasticity in regions without public transportation, while Figure 3.2.3 (2) shows the same for regions with public transportation. As one would notice from (1), the demand curve has a steep slope, and even if the price of gasoline sharply rises, gasoline consumers will simply decrease their consumption levels. Thus, the demand curve here could be said to depict inelasticity. When energy taxes or carbon taxes (that is, environmental taxes imposed on carbon content) equal to t^* on the graph are imposed, the supply curve shifts from S_0 to S_1, the amount that the consumer pays increases from P_0 to P_1 and the price received by the gasoline producer drops from P_0 to P_2. In other words, in areas without public transportation with an inelastic demand curve, the consumers are largely burdened by the imposition of taxes.

Next, in the scenario depicted by (2), where the demand curve is elastic, the difference between the P_0 amount that consumers pay and P_1 is small, so when the price procured by gasoline producers decreases, the difference between P_0 and P_2 is large. In other words, in regions where public transportation is available and demand is elastic, the producers bear a large portion of tax impositions. That is to say, in circumstances such as these, the producers of the goods targeted by the taxes are largely unable to shift the financial burden of the tax on consumers.

Look at the supply curves for Figure 3.2.3 (1) and (2) when their production levels are the same at X_0. The taxes levied are equal to the tax amount multiplied by production levels, or t^* x X^*. Under such circumstances, we also know that to the extent that demand is inelastic, tax revenues will be high. Furthermore, we can see that if we want to reduce pollution to the same degree as could be done in the case of elastic demand, a rather high tax percentage is necessary. Thus, we can see from this how important information about demand elasticity is when environmental taxes are implemented.

With the above example, we considered cases in which the price could be either elastic or inelastic for the same product, gasoline. Generally speaking, however, the price elasticity of demand for energy resources is known to be small. For example, there are debates about how to utilize carbon taxes in Japan.

According to a report by the *Global Warming Tax Expert Committee*, carbon taxation to be imposed on fossil fuels can be divided into two main categories: upstream taxation (e.g. taxes on fossil fuel purification, collections before processing, imports, transaction costs in tax regions) and downstream taxation (e.g. taxes on individual fuel types, purification, movements and wholesales after collections and production plants, direct sales to end users). The differences between upstream and downstream taxation, as well as the resultant affects, can be separated and specified. To begin with, consider upstream taxation.

Japan relies on imports for the majority of its energy, so for manufacturing industries and other energy demand centers of upstream companies, assume that their demand is in an inelastic condition. In this case, the upstream enterprises as the producers of the goods targeted by taxes can shift the carbon tax on to others as the price increases. In other words, it is not safe to say that each part of the domestic upstream enterprises bears the burden of upstream taxes, as the price is shifted, and downstream businesses that use electricity bear the price burden.

Environmental taxes or subsidies?

Certain subsidy policies aim to reduce emissions by appropriating *subsidies* to emitters who successfully reduce their pollution levels. Subsidies are appropriated to reduction amounts that lower production to levels below what is in place at market equilibrium when subsidies are not in place. Below, we will indicate how efficient resource distribution can be realized by using subsidies in place of environmental taxes.

Assume that subsidies equivalent to s^* are allotted to every single unit reduction in production. In other words, when production levels increase marginally, subsidies equivalent to s^* are lost. Accordingly, we could also think of this as an environmental tax worth s^* allotted marginally to production levels. When subsidy funds, instead of a mandate to pay taxes, are to be put in place by the government, it is easy to imagine them having the opposite effect to environmental taxes.

However, that is only the case if subsidies were to pay for increases in production. Since subsidies are paid out when production levels are cut, and thus, when pollution is decreased, it is possible to calculate the optimal subsidy amount in the same way as with environmental taxes.

The subsidy funds, or Pigovian subsidies, necessary to achieve optimal resource distribution must, in the same manner as environmental taxes, be set at the point where production levels are optimized and equivalent to the marginal externality cost (X^* on Figure 3.2.1). Furthermore, even if optimal production levels are achieved through either environmental subsidies or taxes, because the marginal externality cost does not change with optimal production levels, the optimal environmental subsidy (s^*) and tax (t^*) levels are the same by nature.

So, is it safe to say that there is no difference between environmental taxes and subsidies? In the case where subsidies are used and production levels drop from X_M to X^* in Figure 3.2.1, revenues can be considered to have increased in accordance with the reduction levels equivalent to $X_M - X^*$. Yet, in the case of environmental taxes, costs increase in accordance with taxes applied to production levels equivalent to X^*. In other words, compared with environmental taxes, subsidies cause a firm's total costs to become smaller. However, at the same time, this is the same point as costs becoming relatively smaller hand in hand with production decreases. Thus, since environmental taxes and subsidies are set at the same monetary amounts, if production is reduced by the same amount the same cost reductions are also possible. In other words, the difference in the two policies is that subsidies curtail a fixed amount of expenses only. Thus, subsidies are more apt to reduce fixed expenses, and subsequently reduce overall expenses, through increased revenues.

Now, what effect does the difference in such fixed expenses have on the *participation* and *withdrawal* of a firm in the long term? If the firm believes that its profits will increase over the long term, it will participate. However, if the profits were projected to go into a deficit, then it would withdraw from participation in the industry. When environmental taxes are implemented, a firm's fees are raised, so the company's current profits will decrease, and the firms that go into deficit will withdraw from the industry. In this way, the burden of environmental taxes becomes a burden, and companies that go into deficit are those that do not have production methods conducive to environmental conservation. The result of implementing environmental tax is that businesses in certain industries can easily subsist should they adopt production methods that inflict a relatively small burden on the environment. Thereby, industries make the transition having a low environmental burden.

Depending on the industry, its burden on the environment could be relatively large or small. For example, the burden placed on the energy-intensive cement and steel industries is in the instance of carbon taxes, so they decrease in size. Conversely, industries with little negative environmental impact become relatively larger. In other words, the structural makeup of an industry changes as industries make the transition to being less environmentally burdensome.

In the case of subsidies, an opposite result arises. This is because firm costs decrease, so even companies that possess production methods that are largely environmentally harmful can guarantee profits and remain in business. Furthermore, firms that are not acting to change also gain subsidy funds as an added source of revenue, which serves to promote participation in the industry. As more firms participate and the overall number of firms increases, the overall size of the industry grows. Unlike when environmental taxes permit the survival of firms that have a low environmental burden, firms that inflict high environmental burdens are preserved, so harmful firms increase relatively, and firms with low environmental burdens decrease relatively. It is through this process that the potential for great shifts in firm compositions is manifested.

Through what has been described above, when negative externalities arise, we can see that the incentives to participate or withdraw for firms differs in the short vs. long term. Indeed, when participation or withdrawal is not a short-term concern, environmental tax and subsidy systems have the same policy effects. However, when participation or withdrawal becomes a long-term concern, then environmental taxes have the effect of soliciting industries composed of a low environmental burden, whilst subsidies solicit those with heavy environmental burdens. In other words, considering long-term perspectives when implementing environmental policies, subsidies should not be utilized, and taxes are preferable.

Regardless, subsidies have been noted for their efficacy in promoting political objectives in the recent past. As environmental policies, they have often successfully curbed pollution through technological development in a manner that has had profound impacts on local society and industry.

SUMMARY

Environmental taxes are economic measures that aim to curtail pollution emissions by imposing tax fees on them. Environmental subsidies are policies that aim for emissions reductions by awarding subsidy funds to those who successfully cut back their emissions. These two systems have different impacts on firms' short- and long-term decisions to participate or withdraw from an industry.

REVIEW PROBLEMS

1. Explain how to implement an optimal environmental tax.
2. What are the merits and drawbacks of environmental taxes and subsidies?
3. Consider to whom environmental tax burdens are shifted to, and explain how to reverse that shift.

LEARNING POINT: ENVIRONMENTAL TAX AND SUBSIDY PRECEDENTS

In Japan, taxes are mainly applied to fuel, energy, and automobiles. For example, volatile oil taxes are applied to gasoline, light oil transaction taxes are applied to the light oil used by large trucks, and oil gas taxes are applied to the liquefied petroleum gas (LPG) used by automobiles like taxis. Since these suppress energy demand by raising energy prices through tax fees, they are also classified as environmental taxes. However, as the taxes levied are used for road maintenance, coal and oil developments, and other structural support, they are not used for environmental preservation, so they can also be depicted as taxes that do not place the highest priority on environmental preservation.

Regarding subsidies, in Japan, the progress that businesses have made with implementing pollution preventing structures through the present can be measured. Since pollution prevention is achieved in reality, subsidies are promoted through such government-related institutions as the Japan Environment Corporation, the Japan Development Bank, and the Corporation for Small and Medium Enterprise.

Still, there are various ways to measure the forms that carbon and other such environmental taxes take in other countries. Carbon taxes in particular are classified into four main groups in Table 3.2.1. First, there is Sweden's new tax that is imposed upstream. The new carbon tax is imposed on all fossil fuels, and while it did decrease the total existing energy taxes by half, overall, it increased tax amount. Furthermore, fossil fuel wholesalers, producers, and processors are all obliged to pay the taxes, and the collected carbon taxes are pooled into a general fund that is applied to reduce income taxes that were in place.

Next, in Germany, taxes are placed upstream by raising rates on the energy taxes applied to mineral oils (targeting coal and all fossil fuels except for kerosene) and establishing a new electricity tax. The main taxpayers are oil supplying companies (oil mineral tax) and electricity supply utilities. The tax revenues are applied to decreasing the national pension system burdens to both employers and employees.

The UK is an example of a country imposing a new tax downstream. Compared to the hydrocarbon oil tax (which targets gasoline, light oil, jet fuel, heavy oil, and kerosene), this new climate change tax (which targets LPG, coal, natural gas, and electricity) is imposed. Electric utilities and other energy supplying companies are obliged to pay the tax. Additionally, the purpose of the collected tax fees is to return the reduced social insurance burdens of employers back to the industrial sector.

Table 3.2.1 Carbon tax precedents

Classification	Country	Content
Upstream taxation, new tax	Sweden	Imposes new carbon tax on all fossil fuels. Utilizes existing tax collection system to levy fees. Carbon tax allotted as a general fund equal to the amount of reduced income from the allotted income tax reduction.
Upstream taxation, existing tax	Germany	Already existing mineral oil tax rate is raised and a new electricity tax is imposed. Levied taxes are allotted to reduce the National Pension Insurance burden of both employers and employees.
Downstream taxation, new tax	U.K.	New climate change tax imposed on items not targeted by already existing hydrocarbon oil taxes. All levied taxes returned to the industrial sector via reduced social insurance costs and other methods.
Downstream taxation, existing tax	Switzerland	New CO_2 tax imposed on top of already existing mineral oil taxes. Levied taxes returned to citizens and the economy depending upon payment amounts.

Finally, in Switzerland the tax with downstream targets places new carbon taxes on top of already existing mineral oil energy taxes. Moreover, where it is difficult to recognize CO_2 emissions reductions through voluntary initiatives, additional taxes are called for. Switzerland already imports all of its fossil fuels, and taxes are imposed upon their arrival from abroad and transferred to consumers. The levied funds are returned to citizens and industries depending upon the amount paid.

Moreover, while in Europe, carbon taxes have a strong tendency to become general funds, in Japan what is being debated is returning the carbon taxes levied to environmental policies. Current carbon taxes are explained in detail in Chapter 4, Section 4.1.

Section 3.3: Resolutions through direct negotiations

Coase theorem

Usually, solutions to environmental problems are believed to need government or some other sort of third-party intervention. However, this section points out ways by which stakeholders alone can produce solutions through direct transactions and negotiations. More specifically, what are the direct dealings that stakeholders could possible utilize? To elaborate on this, this section turns to a factory wastewater emission precedent.

Assume there is a water reservoir upstream from a house that some people have purchased, and nearby, a new factory has been constructed. The water that the plant emits leads to large increases in the phosphorus and nitrogen in the reservoir, both of which provide nutrients to algae. This leads to the increased breeding of algae in the reservoir, which results in a bad odor that emanates from the lake. At this point, the government could implement the law that sets the standard for the amount of liquid emissions from the plant into the reservoir. In order to abide by the law, the plant would then be coerced to implement liquid waste-processing equipment and reduce phosphorus and nitrogen emissions to levels below legal standards. Furthermore, if environmental taxes were instituted to enforce regulations, tax fees would be levied according to the amounts of phosphorus and nitrogen present within liquid emissions, which would potentially lead to emissions reductions. After calculating and comparing the marginal costs of production and tax fees, the factory would eventually reduce liquid emissions to a certain standard. However, as 1991 Nobel Prize for Economics recipient Ronald Coase purports, even without relying on such problem-resolution methods brought forth by the government, issue stakeholders can use such market mechanisms as voluntary transactions to realize effective problem-solving policies. Depending upon who possesses the water reservoir, related parties can work jointly to bring about solutions.

In other words, Coase stresses the fact that even through such government policies as environmental taxes, market failures can be avoided. This is known as the *Coase theorem*. Issue stakeholders decide on the rights to emit pollutants amongst themselves, and those involved in the pollution can leave it to direct, voluntary negotiations amongst themselves to naturally come to mutual agreements, which leads to the realization of efficient resource distribution. In other words, because the total social welfare can be maximized, government intervention is unnecessary. Thus, in the end, the distribution of privileges, whether they are greater among polluters or those who suffer damages, does not change.

Coase stressed that if the following conditions were fulfilled, inefficiencies such as externalities could be solved through negotiations:

1. Cases where the right for certain entities to use the environment is clearly agreed upon by the whole of society, or in other words, if *rights of possession* are in place.
2. Cases where regulations regarding the responsibility of paying damage compensation are clearly determined.
3. Cases where there are no transaction costs.

Consider, for example, the factory operations in a given region in order to expand upon the idea of conditions conducive to direct negotiations among assailants and victims. One could assume that the noise produced by factory operations is a constant nuisance to the nearby residents. Similarly, the production activities of polluters located upstream of a river yield waste that is unprocessed and dumped into the river, regardless of the harm it would inflict on businesses or residents located downstream. Another example is the relationship between divers and fishermen. Those in the fishing industry have priority access to nautical space for fishing (fishing rights), which has supposedly been the cause of decreased catches among divers. Furthermore, the relationship between smokers and nonsmokers who suffer from secondhand smoke is another pertinent example.

Here, consider the conditions through which both assailants and victims mutually affect each other. First, assume that people have the fundamental right to enjoy a pollution-free environment, otherwise known as a right to the environment. When victims possess such rights, in order for assailants to proceed with pollution activities (e.g. production activities through factory operations or smoke produced from smokers), they must pay damage compensation demanded by victims. To do this, as they pollute, polluters have to calculate the costs they will incur through payouts for the damages in compensation funds or decreases in the utility of victims. On the other hand, when polluters possess the right to pollute, this can be thought of as the permissible right to yield negative externalities. Consider examples where smokers have the right to smoke in smoking sections, and where there are no construction restrictions when houses are being built. In these circumstances, because negative externalities are being produced, social surpluses are not at optimal levels. Therefore, in order to curtail pollution activities, those responsible must pay *compensation*. In that case, the polluters naturally include compensation for pollution reductions into their profit and loss calculations.

The important point of the Coase theorem is that regardless of whether victims or assailants possess environmental rights, the point of ultimately realizing a socially efficient distribution of resource is not lacking in either scenario. With a single glance, it is easy to consider the resultant effect that transactions will have on resource distributions because of how the existence of rights affect subsequent trans-actions. However, as stressed by the Coase theorem, there is no relation between the existence of those rights and optimal resource distribution.

An illustration of the Coase theorem

Scenario where residents have rights

The task at hand is to use a diagram to confirm the effect that differences in the existence of rights have on resource distribution. First, assume that a factory operates in a given locality. Also assume that the pollution emitted by the economic entity performing the production activities (i.e. the factory management) inflicts harm on nearby residents. In this scenario, factory profits increase alongside increases in pollution emissions (and operating hours), so the marginal benefit curve takes the shape of segment FT_1 depicted in Figure 3.3.1. When pollution emissions are to be reduced, factory owners will at first only make reductions to the extent that they do not harm profits, and the costly, difficult to process, emissions reductions will be implemented last. Curve FT_1 has a downward and to the right orientation. As pollution emissions and externality costs conjunctively increase, they inflict harm on the neighboring residents. Curve OG depicts the marginal loss (externality cost) curve. The additional losses that occur alongside emissions increase as pollution levels in a given area increase, so curve OG proceeds up and to the right.

   If there are no environmental regulations in place, the factory managers proceed to maximize profits by expanding operations to T_1 levels, in which case the loss cost

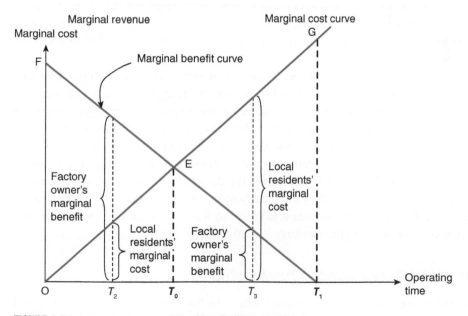

FIGURE 3.3.1 Internalizing externalities through direct negotiations

burdens placed on surrounding residents is not absorbed as a cost to the factory (refer to Chapter 3, Section 1). Thus, operating hours that maximize the social surplus lead to the highest possible profits and eliminate the damages inflicted on local residents. In other words, the socially optimal operating hours are, like the optimal environmental tax found in Chapter 3, Section 3.2, T_0 hours located at intersection E, where the marginal benefit and marginal loss curves intersect. The social surplus here is the area of $\triangle OEF$, or, the difference between the factory managers' profits (equivalent to the area of figure OT_0EF) and the losses incurred by the local residents (equivalent to the area of $\triangle OT_0E$).

On the other hand, the social surplus at T_1 hours is the difference between the factory managers' profits (equivalent to $\triangle OT_1F$) and the losses incurred by local residents (equivalent to $\triangle OT_1G$), or equivalent to duplicate $\triangle OET_1$, so they become $\triangle OEF -\triangle EGT_1$. In this case, the local residents suffer enormous damages, which must be reduced by $\triangle EGT_1$. Thus, if there are no regulations in place, the damages to society equivalent to $\triangle EGT_1$ arise in comparison to those that come about during optimal operating hours.

Now, assume that local residents have a right to enjoy a clean natural environment. If they exercise that right to their fullest ability, factory operating hours will be forced to drop to the origin point O and emissions amounts will be zero. Accordingly, in order for the factory to do production activity, the negotiation point starts from origin point O, and the residents must approve the operating hours. From this premise, both parties begin negotiations, and if they can agree on compensation amounts for the damages the factories place on the local environment and its residents, then operating hours can be increased. Since profits earned by the factory become equivalent to OF as the damage inflicted on local residents is 0 when operating hours increase from 0 to 1, the factory employs compensation to make up for the local residents' losses. Next, when considering operating hours that increase to T_2, the benefits to the factory surpass the marginal losses of the local residents, providing further incentive to increase operating hours.

In this way, the factory that increases compensation does not add on profits, when at operating hours of T_0, in other words, where the marginal benefit and marginal losses are the same. For instance, considering the scenario when operating hours are increased to T_3 from T_0, the marginal benefits to factory owners are below the marginal losses to local residents, and there are no incentives to increase operating hours. Thus, negotiations between both parties break down, resulting in a return to point E. The area of $\triangle OT_0E$ that represents the damage costs inflicted on residents is exchanged by the amount paid by the factory to the residents, making T_0 hours allowable. Thus, the remaining amounts equivalent to the area of $\triangle OEF$ after the factory has paid compensation to the residents is retained as profit.

Scenario where factories have rights

Conversely, we can imagine scenarios where factories originally have rights. For example, consider the scenario where a factory that has, for some time, been operating in the middle of the mountains where there had been absolutely no residential development. Then, all of a sudden, the area around the factory is to become developed into a residential area. In this case, the starting point of negotiations is point T_1. At the start, the factory earns profits equivalent to the area of $\triangle OT_1F$, and the residents incur losses equivalent to the area of $\triangle OT_1G$. In this case, residents have incentive to negotiate with the factories to get them to decrease their operating hours to points below T_1. The reason for this is that marginal decreases in operating hours from T_1 do not largely decrease profits, but the local residents can avoid damages equivalent to GT_1. Next, if one were to consider a decrease in operating hours from T_1 to the lesser T_3, it is clear that the marginal losses of the local residents surpass the marginal benefits of the factory owners, so there is an incentive to drop the operating hours. Through negotiations of these aims, compared to the diminishing profits that the factory suffers as operating hours are curtailed, they will be continued to the extent that the loss evasion amount of the local residents is small. Therefore, the residents pay compensation to the factory for its losses in marginal benefit in order to get them to cut back their operating hours. Through this, the most efficient operating hours for society are realized.

The results of the above negotiations are that local residents pay an amount equivalent to the area of ET_0T_1 in compensation. The terms of adopting this compensation is that the factory agrees to curtail operating hours down to T_0 levels. Even by doing this, the factory gains profits equivalent to the area of $\triangle OT_1F$ through compensation increases in addition to profits equivalent to the area of space OT_0EF when operating hours decrease to T_0 levels. Even if subsidies are paid out, the local residents' losses go down to an area equivalent to $\triangle EGT_1$ at T_1 hours, so local residents' economic conditions improve. In other words, regardless of whether or not the environmental rights belong to those harmed or the not, or whether or not the right to pollute are owned by polluters, the operating hours that determine the final efficient redistribution of resources do not change, and they become equal to T_0.

Thinking about it like this, if the rights to possessing benefits and damages borne out of environmental externalities are clearly determined at the outset, and regardless of who has the rights, a local factory and residents can, through voluntary negotiations, achieve a distribution of resources that is most efficient to society. This is the essence of the Coase theorem.

Problems with the Coase theorem

However, in reality, there are many cases where parties with a vested interest cannot smoothly solve externalities as depicted in the Coase theorem. Table 3.3.1 depicts a number of issues inherent in Coase theorem assumptions.

The first issue is with regards to the influence attributed by changes in rights distributions. If the distribution of rights changes, the most efficient resource distribution for society becomes achievable based upon new marginal benefit and marginal loss curves via methods that redistribute possession rights, and the resource distribution is different than what was in place before the two curves changed. In reality, rights distribution is considered to largely affect profit and income distributions among the

Table 3.3.1 Issues with the Coase theorem

Issue area	Description
Impact on rights distribution	The Coase theorem presupposes that the marginal benefit curve of polluters and the marginal damage curve of victims do not change based on changes in rights distributions.
Transaction costs	The Coase theorem assumes that consensus is reached and enforced without inconveniences during negotiations between parties of vested interest.
Imperfect information	When there is imperfect information regarding the marginal benefits and damages of those involved in negotiations, efficient negotiations are not possible.

polluters and victims related to the externalities. For example, when neighboring residents possess rights to the environment and could potentially increase their income (via compensation) through negotiations, it is possible for the marginal losses curve to go up. In this case, in addition to income increases, the local residents demand improved environmental quality. In contrast, real incomes for factories decrease due to compensation payouts to neighboring residents. Because of these real income decreases, it is possible for the marginal benefits of the polluters to decrease as well. Thus, the intersection of the two curves changes, and as a result of negotiations, the actual hours of operation do not amount to the T_0 depicted on the graph.

The second issue is the size of transaction costs. In order for the Coase theorem to be valid, related parties must be able to easily negotiate. However, in reality, one could imagine a plethora of scenarios where transaction costs would be quite high. For example, the organization that possesses pertinent rights could have challenges with regulations. There would be no issue there were only one local resident and one factory manager. However, if there are many constituents, everyone's rights must be determined ahead of time, and all of the stakeholders related to such regulation must be recognized.

For example, during negotiations, where someone serves as a proxy for those who live nearby, it is difficult to specify who that person stands for. Furthermore, for people who are separated by a significant distance from the factory, it is difficult to specify the losses that they incur. As we learned in Chapter 2, Section 2.3, environmental problems and the losses that they generate are, like public goods in nature, difficult to articulate along the lines of price. For instance, one could speculate the degree to which health damages are incurred from air pollution, but it is difficult to speculate the monetary price for emotional unpleasantness. Furthermore, in dense factory zones, how the responsibility for pollution control should be allocated becomes an issue. In this way, when there are many entities involved in negotiations, they become difficult to carry out.

In other words, overseeing and managing who acquires what rights and to what degree and from whom through negotiations is difficult to determine. Because of this, in order to carry out negotiations, transaction costs arise. Moreover, to the degree that there are many people involved in the negotiations, the free-rider problem in which a person or people choose not to participate in negotiations because someone else will accept responsibility for more than what they are accountable for could arise. And of course, there is also the possibility of the worst-case scenario in which negotiations do not occur at all. In other words, if negotiation costs are quite large, then voluntary negotiations alone do not produce efficient resource distribution.

The third issue is the problem of imperfect information. For example, when local residents are recognized as possessing the rights to the environment, they will declare that the losses that they incur are large and work to draw out as much compensation money as possible from factories. It is not difficult to imagine that each party would operate strategically to realize resource distributions that are comparatively favorable to them, so negotiations would not conclude smoothly, and large transaction costs would arise.

Even Coase himself was not aiming to indicate that externalities could be internalized through the voluntary negotiations depicted in the theorem. The aim of the Coase theorem was not to encourage unilateral engagements, but to use negotiations to address free riding and other issues that arise not from oneself, but from the inappropriate behavior of others. Overall, the theorem incites efficient resource distributions and highlights the importance of the existence of transaction costs.

There are many transaction costs, including costs of specifying parties with vested interests, costs of making space for negotiations among relevant parties, the income foregone by residents upon participating in negotiations, and the necessary time costs of strategic negotiations. Here, it is essential to think about the various benefits and transaction costs as efficient resource distribution is put into place and residents' utility increases after negotiations. In this case, transaction costs for negotiations are enormous, so transaction costs of negotiations can be thought of as larger than the benefits. This could also be called a condition in which not performing transactions would be efficient. Thus, when negotiations among relevant persons do not lead to the internalization of externalities, it could become necessary for the government or justice departments to intervene.

Issues regarding environmental rights and resource distributions

With the Coase theorem, when relevant parties can negotiate resource distribution without incurring costs, the achieved pollution levels will be optimal and the same regardless of whether or not the victims possess rights to the environment or polluters possess rights to pollute. In other words, regardless of who possesses rights, negotiations solve problems, and social benefits are maximized. Accordingly, looking at this from the perspective of efficient resource use in society, environmental organizations and nature preservation activity have no persuasive power in their emphasis that they possess environmental rights.

However, deciding rights possession has a great impact on the distribution issue with regards to which subsidies are received. Here, as an example of a pollution problem, consider the Japanese pollution trial where the business that caused the pollution had to pay compensation to those harmed. Specifically, look into why it was appropriate to allot environmental rights to the victims.

First, local residents who worry about pollution are not necessarily part of the wealthy class. And, when injuries are inflicted upon great numbers of people, the financial compensation deducted from polluters' profits and attributed to victims is high. Therefore, the decrease in costs yielded from the victims seeing the villains decreases, so the government set the *Pollution-Related Health Damage Compensation Act* in 1973. Specifically, this deals with the financial burden that businesses and factories across the country must bear by specifying how much one meter cubed of sulfur dioxide emissions costs. The government, thus, played a one-time role in lowering the transaction costs of emissions between polluters and victims. In this way, when it is difficult to specify actual rights possessed, the government or judiciary can extend its hand into initial problem-resolution proceedings, and after that, negotiations can be entrusted to relevant parties.

SUMMARY

The Coase theorem states that when there are no transaction costs for negotiation, and the possession of rights and compensation payment obligations are predetermined, negotiations allow for solutions that maximize overall social welfare. This, it goes on to say, is true regardless of whether victims or assailants are responsible for paying compensation. However, there are many cases in which stakeholders are unable to resolve externalities as smoothly as the manner depicted by the Coase theorem since it presupposes that there are no transaction costs and that there is perfect competition.

REVIEW PROBLEMS

1.   Outline Coase theorem.
2.   Why are transaction costs important?
3.   What are the limits of the Coase theorem?

LEARNING POINT: THE REALITY OF MULTIPLE TRANSACTION COSTS

Especially in Japan in previous years, there was great potential for the transaction costs inherent in negotiations to grow quite large. For example, the atmospheric pollution brought about from 1960 to 1972 in Yokkaichi City, Mie Prefecture led to widespread outbreaks of asthma, and litigation for Yokkaichi asthma (one of the four major pollution-caused diseases in Japan) cases took over ten years to settle.

In this way, when a lot of time is involved from the time a case is presented in court until the time judgments are made, large transaction costs can be expected.

Moreover, in 1988, 472 plaintiffs, all of whom were residents of Amazaki City, charged the government, the Hanshin Expressway Public Corporation, and nine other firms as the harmful polluters in the Amazaki air pollution suit. The courts decided the government's responsibility in the affair in the year 2000, but it was also difficult to specify negotiators. The suit was brought forth because of automobile exhaust and factory smoke, and the polluters were factories and automobile drivers. It was difficult to estimate and identify the time and location information of each individual driver. Furthermore, since negotiation partners related to individual automobile drivers could not be determined, it became clear that the Coase theorem did not apply in this case.

Another precedent case is climate change induced by greenhouse gases yielded from fossil fuel energy consumption (refer to Chapter 1, Section 1.3 for details). The perpetrators (i.e. polluters) are people from current and past generations, while the victims are people in future generations. In this scenario, it is impossible to hold negotiations between the current and past generation perpetrators and future generation victims, and as policy absolutely fails to mediate between the two groups, this issue is not applicable to the Coase theorem. In other words, policy that can serve as a mediator is essential.

Beyond negotiations among related parties, through markets created by the government with the establishment of tradable greenhouse gas emissions rights, every country's overall pollution emissions should diminish as transaction costs are abated. Furthermore, companies can be directed to only emit what they are entitled to via greenhouse gas emissions privileges stipulated by the government. And, if a company has excess emissions rights (e.g. they did not emit as much as they could have), they can sell the remaining rights to other companies, and companies that need to emit more can purchase the rights from other companies that have them available. In this case, the government has a duty to oversee emissions frameworks and levels (the specifics of emissions trading are outlined in Chapter 3, Section 3.4).

With emissions trading, trading does not occur between those inflicting and those experiencing harm, but among those who cause harm exclusively, thereby allowing the problem of transaction costs through negotiations to be eliminated. Through this, emissions reductions can be implemented at the lowest possible costs. However, in order for these transactions to occur smoothly, the government must aid in creating markets.

Section 3.4: Emissions trading

What is emissions trading?

In the previous section, attention was paid to the financial rights of environmental assets with the Coase theorem. In this section, the adopted *emissions trading* (also known as emissions rights trading and emissions certificate trading) systems are applied to the Coase theorem. Businesses, factories, and other related entities are granted emissions certificates for a standardized amount of pollution output. In many cases, the government introduces emissions permission certificates that give companies the right to pollute up to certain levels. Then, a market for trading such emission certificates among companies and organizations that hold them is created.

Polluters have to possess emission certificates that exist in the trading market, and if each business or factory or other organization pollutes more than the permissible amounts outlined in their emission certificates, they can buy additional emission rights from other polluters. Conversely, if they pollute less than the amount permitted by their certificates, they could sell their excess certificates. In this way, this is a system that can measure overall emission reductions efficiently while also reducing transaction costs. This is accomplished not through negotiations among factories and businesses, but through the trading of emissions certificates allocated by the government. And, when other economic methods are conjunctively utilized to construe monetary value to environmental assets in order for them to be taken in by the market, externalities can actually be internalized.

Emissions certificates determine the marginal unit denominations of pollution, and those who possess such certificates are permitted to pollute as much as the unit specified on the certificate. Polluters can emit as much as they would like so long as it is within the permissible amounts specified by the emissions certificates.

However, the total permissible amounts of emissions for all of society are determined beforehand by the government. Aggregate emissions levels are overseen by the government (the emission trading system's administrative authority), and they are suppressed according to how many emissions certificates are issued. Regulating overall emissions in this way is one of the advantages of emissions trading systems.

Via sequential decreases in overall permissible emissions amounts, it is possible to greatly decrease overall emissions amounts over the course of multiple years. For instance, between 1990 and 1995, the U.S. Environmental Protection Agency successfully reduced SO_2 emissions from 8.7 million tons to 5.3 million tons through targeted regulations of this sort.

When pollution emission amounts can be freely traded and transferred in the market, anyone participating in the emission market can trade emission amounts by buying and selling emissions certificates. In other words, by implementing emission trading, those who can easily decrease emissions can do so, and those who cannot easily do so can acquire the right to emit the amounts that they need. Through this, just as depicted in the following graphs, it is possible to exert an equal burden (marginal cost) on all businesses for decreasing emissions, and since businesses can optimize their production activities amongst each other, the social cost of implementing reduction targets decreed by the government are minimized. However, as indicated by the Coase theorem, optimal emissions amounts are not dependent upon how they are apportioned at the start. The purchase and sale of pollution emissions privileges through emissions trading schemes is, despite having sounded bad previously, a reasonable system in reality.

An illustration of emissions trading

Figure 3.4.1 illustrates how emissions trading systems function. Here, a company invests resources and produces marginal goods simultaneously as it emits pollutants. The curve that proceeds downwards and to the right represents the business' marginal benefit for pollution emissions. When emissions are free of charge and the company can emit as liberally as it pleases, the total emissions will reach their maximum levels, as depicted by point X_M. This is where the market price is 0.

Here, the horizontal straight line P^* on the graph indicates the market price. In other words, for every unit of pollution emitted, the company must pay price P^*. For the company, X^* units, which arise at the point where the marginal price P^* and marginal benefit curves intersect, is the optimal emissions levels where profits are maximized. The reason for this is because at pollution levels greater than X^*, the costs of emissions surpass marginal benefits, and the more emissions levels increase, the more profits decrease. At any emissions level, if the market price is set, we know from the Coase theorem that the optimal emissions levels for the company are dependent upon the costs of emissions. Thus, it is not a matter of how much was being emitted initially. We will explain this with two examples below.

First, assume initial emission amounts equivalent to X_1. In this scenario, the company will only purchase emissions supplements equivalent to $(X^* - X_1)$ from the market. However, if the initial emissions amounts were at X_2 levels, the company would have excess emission amounts equivalent to $(X_2 - X^*)$. Thus, this company could sell the excess emissions amounts to the markets to gain additional profits.

Next, in order to consider resource distribution within an emissions trading market, first, consider the relationship between the marginal abatement cost and the marginal

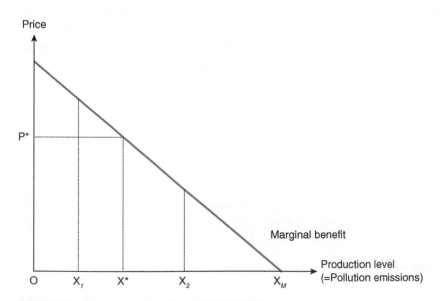

FIGURE 3.4.1 Emissions trading of a single enterprise

benefit. According to Figure 3.4.1, the input amount is determined by the company's marginal benefit, which is equivalent to emissions cost P^*. Thus, in scenarios where emissions were, conversely, decreased, the marginal benefit indicates how much the company profits decrease per marginal decrease in emissions amounts, so this can be interpreted as the marginal abatement cost.

From here, horizontally invert Figure 3.4.1 in order to analyze the marginal abatement cost (Figure 3.4.2). If the emissions cost is 0, then emissions amounts are X_M. If emissions amounts are decreased from point X_M, then emissions reduction amounts move by that amount to the right. When emissions prices surpass marginal reduction costs, increasing emissions reductions lead to additional profits, so with regards to emissions price P^*, the optimal emissions reduction are equivalent to $(X^* - X_M)$.

Combining Figures 3.4.1 and 3.4.2 yields Figure 3.4.3. On this graph, point O_1 is one company's emissions origin, while point O_2 is another company's emissions origin. If there are no emissions regulations, company 1 will emit as much as $O_1 S_M$, and company two will emit as much as $O_2 X_M$. In other words, the length of line segment $O_1 O_2$ indicates the overall emissions amounts for the whole of society in scenarios where there is no regulation in place.

Next, take a look at Figure 3.4.4 in order to consider society's standardized reduction levels. In order for the government to reduce emissions amounts equivalent to $O_2 O'_2$, it distributes traceable emissions amounts equivalent to the length of $O_1 O'_2$ to both companies. In order to decrease pollution levels by $O'_2 O_2$, Company 2 shifts from its origin point O_2 to point O'_2, resulting in a supply curve shift to the left. In Figure 1.3.4, Company 1's initial emissions amounts are equal to $O_1 A$, while Company 2's are only worth AO'_2. If emissions trading did not take place, then both companies would decrease emissions amounts from their initial amounts, meaning that Companies 1 and 2 realize marginal benefits of equivalent to AC and AB, respectively. Compared to Company 2, Company 1's marginal abatement cost is high. Therefore, since the fee to decrease emissions is so great, there is added incentive for both companies to work together through emissions trading.

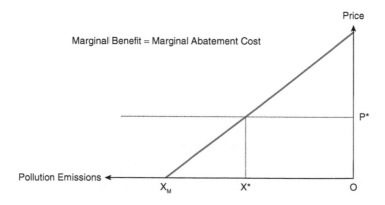

FIGURE 3.4.2 Interpreting the marginal externality cost from the marginal benefit

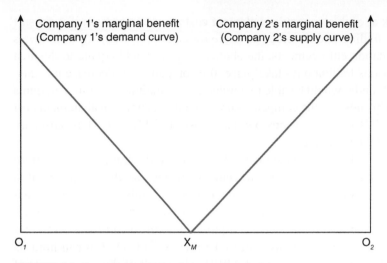

FIGURE 3.4.3 Demand curve in the absence of regulations

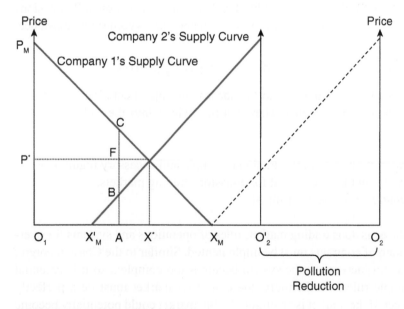

FIGURE 3.4.4 Optimal emissions level

　　Here, through emissions trading, Company 1 will buy emissions amounts from Company 2. The marginal benefit curve of the purchasing Company 1 indicates the demand curve, and the marginal benefit curve of the vending Company 2 indicates the supply curve. Furthermore, the emissions trading price $P*$ is where the supply and demand are equal, and the amount of emissions traded is equal to $AX*$. This, at point D where market equilibrium is realized, realizes a more efficient resource distribution compared to cases when the government restricts overall emissions.

Next, take a look at each company's abatement costs. For company 1, if it doesn't perform emissions trading and is forced to decrease emissions from X_M to A levels through direct regulation enforcements, the abatement cost would equate to the area of $\Delta X_M CA$. If emissions trading does take place, the company will decrease its emissions from X_M to X^*, and it would be able to purchase the remaining emission amounts (e.g. emissions certificates), so the company's costs would become the abatement cost equivalent to $\Delta X_M X^* D$ + the emissions purchase cost of $\Box X^* AFD$. Thus, costs are only restricted by ΔDFC amounts.

On the other hand, in the case of Company 2, in order to decrease emissions from X_M to A levels through direct regulation enforcements, its fees would be equivalent to the area of $\Delta X'_M BA$. If emissions trading occurs, emissions equivalent to AX^* (emissions certificates) will be sold, and emissions amounts would drop from X'_M to X^*. The company's abatement costs will become equivalent to the area of $\Delta X'_M DX^*$, but as it gains emissions sales profits equivalent to the area of $\Box AFDX^*$, it can limit its expenses to amounts only worth the area of ΔBFD. The result of this is, the costs of both companies are restricted to a mere ΔBCD through emissions trading. In this way, based on the market equilibrium realized by this distribution, when overall emissions allotments are set, a harmony in the surpluses of both companies comes into existence.

Obstacles to adopting emissions trading systems

There are many conditions conducive to the successful implementation and coordination of emissions trading systems. They can be divided into the following three sub-categories:

1. *Market creation* through the introduction of simple and necessary regulations.
2. *Perfect competition* that is essential for emissions trading markets.
3. *Initial distribution* must be determined.

First, within the emissions trading market, rules of operations and systems for overseeing the execution of contracts must be implemented. Similar to the Coase theorem, transaction costs will increase as the system becomes too complex, so it is essential to implement simple rules. Furthermore, the emissions market must be a perfectly competitive market. If the market is an oligopoly, the market could potentially become controlled. For instance, if there are a few participating enterprises, they could buy up all of the emissions rights and prevent the sale of such rights to newcomers in the market, effectively preventing entrance into the market.

Two representative methods for initial distribution are competitive bidding and *grandfathering*. Since the government receives revenue from auctions when competitive bidding is held, companies that already pollute do not wish for systems to be implemented through auctions. Rather, they would prefer an emissions trading system in which initial emission amounts are distributed free of charge. If it is independent of the size of the company, every company can be allotted the same privileges for free. If trading occurs under such circumstances, the emissions amounts over the implementation years will become as great as those emitted by the largest polluters up until then, leading to great damages. On the contrary, to the extent that a company does not pollute, there are revenues that could be gained from selling emissions rights, leading to profit gains. In other words, this kind of distribution is thought to be difficult to pull off. Yet, there is also a system known as grandfathering in which a proportion of the previously possessed emissions rights are already redistributed for free.

Grandfathering is often criticized for not attributing financial burdens upon review of negative externalities brought about by emissions. However, the alternative competitive bidding method is thought to be entwined with the difficulty of achieving perfect competition, and the grandfathering method is said to be ideal in that it does not extremely and unevenly distribute initial emissions levels as it is based on an appropriate criterion such as the establishing an equity of burdens.

LEARNING POINT: EMISSIONS TRADING PRECEDENTS

For quite some time now, emissions trading has been used in a number of environmental policies in countries throughout the world. They are pertinent not just for air pollution, but also for land management, water resource management, and fishery management. For example, emissions trading was one aspect of air pollution policy in the U.S. during the latter half of the 1970s. From the lead trading system implemented in the oil purification industry in the 1980s, to the sulfur dioxide (SO_2) and nitrogen oxide emissions certificate trading system 1990s, emissions trading systems have been utilized in numerous actual environmental policies. Among them, the sulfur dioxide emissions certificate trading system is famously known as a successful example. There are ongoing investigations into the extent to which the costs vary between when direct regulations are enforced and emissions trading is permitted. The fact that cost reductions that arise from trading increase with time is stressed in particular.

Throughout Australia, America, and Chile, there are two major differences in scope of application for emissions trading systems related to water. The first is trading the rights to draw in water or emit used water in order to manage water resources. The second targets the preservation and management of surface water through emission level or water pollution trading.

Individual transferrable quotas (ITQs) have been applied in Australia, Canada, Iceland, the Netherlands, New Zealand, and the U.S. for fishery management (refer to the **Learning point** in Chapter 2, Section 2.2 for an explanation of ITQs). Traditionally, fishery resources are managed based on the idea that the right to fish is equally divided among everyone. If everyone is entitled to fish in the future, the result is that many do not think to preserve fishing resources and fish stocks diminish. Therefore, permissible catch loads are allotted to every person who fishes, and the management of these catch loads is managed through a rights trading market, leading to stable and consistent catch loads.

AN ECONOMIC EXPERIMENT WITH EMISSIONS TRADING

The following experiment can be performed to gain a deeper understanding of emissions trading. It can be found online at this book's study support corner website. Check out the details at the following URL:

http://homepage1.nifty.com/kkuri/research/EnvEconTextKM/

SUMMARY

In emissions trading systems, the right to emit single units of pollution is standardized, and the rights to trade those rights in a market are established. Accordingly, if the government determines in advance the target emission amount standards, then the total economy can restrict pollution emissions to levels within target levels.

REVIEW PROBLEMS

1. Describe emissions trading systems.
2. How could emissions trading systems potentially lead to efficient resource distribution?
3. What are the problems with initial distributions in emissions trading systems?

CHAPTER 4

Applications for environmental policies

Chapter overview

The central focus of this chapter is the ideology behind environmental policies. The first topic to explore is the means by which appropriate policy measures are chosen out of a number of alternatives. While economic measures are thought to be superior to direct regulations for achieving environmental policy targets in a cost-efficient manner, many policy decision precedents point to why many countries have traditionally employed direct regulations. Next, this chapter highlights the recently predominant yet heavily debated environmental tax and emissions trading systems. Following this is a depiction of how such policy measures as direct regulations and economic methods can be efficiently combined. The chapter concludes with considerations about important aspects related to future economic theory by honing in on waste problems in Japan as well as on the climate change issue that will affect both developed and developing countries.

Chapter content

Section 4.1—Issues with decision-making within environmental policy measures are of considerable significance in environmental economics. This section explains how to combine direct negotiations, economic measures, and other processes, while detailing the types of policy measures that are ultimately chosen.

Section 4.2—This section introduces the types of procedures involved in the waste disposal fee and deposit systems, both of which impose cost burdens according to the type of waste being disposed. Then, how each of these methods is appropriate for certain conditions is explained.

Section 4.3—Nowadays, measures to address climate change are being debated across the globe. This section introduces the Kyoto Protocol and its market-based mechanisms.

Section 4.4—When it was enacted, the Kyoto Protocol was relatively ineffective for preventing climate change. To date, the global community attempts to determine which policies could effectively succeed the Kyoto Protocol. This section explains how efficiency ought to be inherent in future policies.

Section 4.1: Policy choices

Problems with choosing environmental policies

Adopting particular environmental policies requires a series of debates and other processes. Policy issues must first be recognized and consented upon. Following this, appropriate issue resolution and objective achievement must be approached through policy planning, proposal, deliberation, determination, implementation, evaluation, and finally, updating. Of prime importance to these many processes is understanding how certain policy methods are comparatively more or less efficient within a particular set of circumstances. The main aim of this section is to assess the unique and special features of various policy processes.

Decision-making issues within policy processes are important topics in environmental economics. As we have previously seen,

1. Economic methods are better at bringing about cost efficiency than direct regulations, and
2. Among economic methods, long-term perspectives about participation and withdrawal, as well as perspectives about the equality of cost burdens, demonstrate that environmental taxes and emissions trade systems are superior to subsidies.

Moreover, the reason that environmental taxes and emissions trading systems yield the same effects is that tax rates determine emissions standards in the case of environmental taxes, and certificates set the standards within emissions trading systems.

This section will proceed by first comparing technology regulations and economic methods among direct regulations and explain why many countries have traditionally employed them. Following this, a comparison between the policy measures involved in environmental taxes and emissions trading systems will be provided. The final part of this section will serve to explain effective policy measure combinations.

Technology regulation and economic techniques

The shortcomings among emissions' standard direct regulations were outlined in the previous chapter. The next step is to expand on this with a focus on direct regulations that are different in nature. Consider, then, regulations on business operating conditions, such as those that require firms to adopt certain manufacturing technologies or install pollution-prevention instruments. For instance, air and liquid emissions standards in Western countries are based on the processing levels implemented through the "Best Available Technology Economically Achievable" standards. As another example, the U.S. Musky Law aimed to prevent air pollution by requiring certain technologies to be standardized in automobiles. And, currently in Japan, domestic regulations on automobile emissions are enforced alongside construction permits and other controls to ensure the import and sale of automobiles that meet determined exhaust concentration standards only.

As pointed out in the previous chapter, direct regulations differ from economic methods and are not necessarily efficient. The reason for this is that first of all, production-induced pollution emissions cannot be regulated when certain special technologies are compulsory. Moreover, since companies must only equip designated technologies in order to adhere to the system, they have no incentives to develop more environmentally friendly technologies and cut emissions beyond what is required. This further translates into few adequate incentives to invest in technological research and development. Finally, when technological standards are determined, it is then difficult to provide the funds to incorporate further, more efficient environmental technologies. With all of these considered, direct regulations are often cited as ineffective.

Environmental taxes and emissions trading systems

Previous chapters discussed the information that governments possess that allows or prevents them from properly grasping resource supply and demand. In order to implement optimal environmental policy, governments require information related to both profit downturns (the marginal benefit curve) driven by production downturns and social costs (marginal externality costs) that arise from pollution emissions. Assumptions about the existence of perfect information, or information that is complete and free, have allowed environmental tax and emissions trading systems to be considered and compared with relative ease. However, in reality, there is a lot of information that is difficult for policy authorities to acquire properly without a price.

With perfect information, resource distributions within environmental tax and emissions trading systems achieve the same results. However, governments cannot always acquire sufficient information about environmental issues like climate change when drafting relevant policies. And even when information is attainable, environmental changes, technological developments, and other *uncertainties* come into play. The following paragraphs explain the differences between instances of uncertainty among environmental tax and emissions trading systems.

First, uncertainty in the context of environmental taxes renders it difficult to identify and set an optimal tax rate. It is thus determined through trial and error, leaving no other choice but to tinker with emissions standards and gradually approach the optimal tax rate. Another way of thinking about this is that determined tax rates, regardless of whether or not they are derived through trial and error, can be adjusted through policy, yet pollution standards-related uncertainty remains unavoidable. Furthermore, unlike consumption taxes where rates are fixed proportions of sales prices, when taxes are applied to marginal units of pollution (e.g. through environmental taxes), actual environmental tax rates fluctuate as commodity price levels fluctuate at the whims of inflation and deflation. These fluctuations may also lead to changing pollution standards.

On the other hand, when there is uncertainty in emissions trading systems, optimal total emissions levels are not concrete. Even if overall emission levels are standardized to some degree, policy authorities are unable to predict price fluctuations since the market determines emissions prices. In other words, uncertainty renders ideal prices ambiguous, market participants are compelled to strategize and speculate, and emissions permission certificate prices vary wildly.

After due consideration of all that has been previously stated about the two systems, it is evident that there are a number of pros and cons for each, and as such, it is unwise to take a "one-size-fits-all" approach to regulatory frameworks. If that is so, then, what should government policy decision-making be based upon? Beyond the existence of uncertainty, many believe that the system that inflicts the least damage should be adopted since at least some unavoidable negative aspects arise from both. Judgments from this sort of "minimal damage" perspective require analyses of the comparative slopes of the marginal benefit and marginal externality cost curves, which are to be discussed at length in subsequent sections.

The effects of uncertainty

Figure 4.1.1 displays various political processes to be employed when uncertainty exists and governments cannot acquire sufficient information regarding the relationship between the marginal benefit and marginal externality cost curves. Just as in the previous chapter, the prices are measured by the vertical axis, pollution that arises from commodity production levels are measured by the horizontal axis, MB depicts the marginal benefit curve, and MEC depicts the marginal externality cost curve as emissions rise. Additionally, point e is where the aforementioned curves intersect, production levels reach X^*, and net social gains are maximized with fixed production-induced emissions.

Marginal benefit curve information uncertainty

Imagine a case of uncertainty where the government cannot acquire accurate information about the marginal benefit curve MB. Consider whether the policy authority should roughly assume the MB curve to be the same as MB_F and impose tax rates t^*, or whether it should determine a regulatory framework that hinges production levels

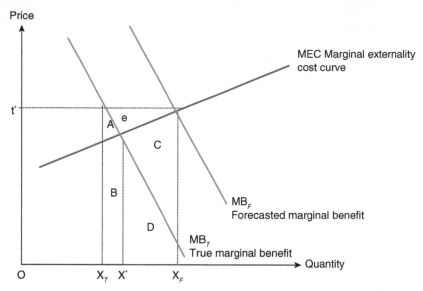

FIGURE 4.1.1 Uncertainty surrounding marginal benefits (Case 1)

to X_F via an emissions trading system. Also assume that MB_T represents the appropriate marginal. In this scenario, adopting the policy based upon the incorrect assumptions about the MB curve would undoubtedly give rise to social losses. Yet, were environmental taxes to be implemented, businesses would decrease production levels to X_T, where the proper marginal benefit curve MB_T and the tax rate t^* are equivalent. The results of this are that, compared to scenarios with optimal X^* production levels, profits decrease by the area of $(A + B)$, yet externality costs only decrease by the area of B through emissions reductions. The damages incurred by the tax system thus become equivalent to area of A for both parties combined.

   On the other hand, since the regulatory framework of the emissions trading system drives production levels to X_F, production greatly rises from D^* to X_F. The additional, marginal social externality cost incurred is equivalent to area $(C + D)$, and since profits only increase by area D, the harm inflicted on society becomes C. At this point, take another look at Figure 4.1.1, and recall the hypothesis that the slope of the marginal benefit is greater than the slope of the marginal externality curve. Therefore, the area of A is less than the area of C. Environmental taxes should thus be adopted in this case since they inflict less social harm.

   However, in the opposite case where the slope of the marginal benefit curve is less than that of the marginal externality cost curve (as depicted in Figure 4.1.2), the area of A is greater than area C, so an emissions trading system would inflict less social harm. In this way, government decisions for adopting environmental taxes versus emissions trading systems are dependent upon the comparative slopes of the marginal benefit and marginal externality cost curves. When this evaluation method is applied to climate change, which policy measure turns out to be best?

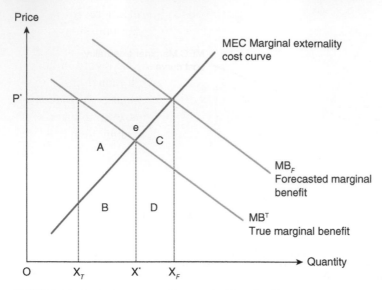

FIGURE 4.1.2 Uncertainty surrounding marginal benefits (Case 2)

The concentration of climate change-inducing greenhouse gases is seldom a simple single year measurement of emissions, but rather a value that includes cumulative, year-to-year emissions levels. For gases such as CO_2, nitrous oxide, and furon that respectively remain in the atmosphere for 200, 114, and 45–200 years at a time, even if their future emissions were successfully curbed, there would be no additional benefits. This means that the shape of the marginal externality cost curve would become nearly horizontal, rendering the slope of the marginal benefit curve comparatively larger and rendering environmental tax systems ideal. Based on this logic, environmental taxes are now prioritized among climate change policies. Yet by the same logic, when harmful emission levels are curbed and produce the opposite comparative relationship between the marginal externality cost and marginal benefit curves, emissions trading systems are ideal.

Marginal externality cost information uncertainty

What happens in situations like the one depicted in Figure 4.1.3, when there is uncertainty not with the marginal benefit, but with the marginal externality cost curve? The underlying logic is the same as that of Figure 4.1.1. From the start, the policy authority must decide which policy measure it wants to utilize, be it imposing tax rate t^* based on the mistaken assumption that the marginal externality cost is placed along the MEC_F curve, or establishing the regulatory framework for an emissions trading system with fixed X_F production levels. However, MEC_T depicts the actual marginal externality cost. Thus, through mistaken estimations, social harm is incurred. Where environmental tax policies are in place, companies endeavor to decrease production

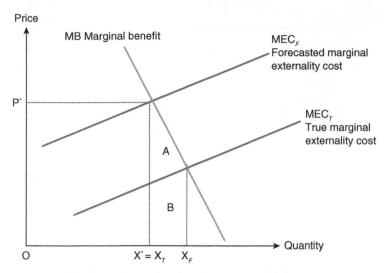

FIGURE 4.1.3 Uncertainty surrounding marginal externality costs

levels to X_T, where the tax rate $t*$ is equivalent to the marginal benefit curve MB. The result of this is that, compared to scenarios with optimal $X*$ production levels, profits only decrease by amounts equivalent to the area of $(A + B)$, yet externality costs only decrease by the area B since emissions reductions are so small. With all factors considered collectively, the environmental tax policies render social harm equivalent to area A.

Conversely, since the regulatory framework of emissions trading systems drive production levels to X_F, production drops from $X*$ to X_F. At that point, reductions in externality costs amount to area B, profit shrinks by the area of $(A + B)$, and overall social harm equates to area A. What is clear from the above is that regardless of the policy measure, social harm will not change. There is thus no preferable system when there is uncertainty regarding the marginal externality cost. Therefore, only cases of uncertainty with the marginal benefit are considered when choosing between environmental tax and emissions trading systems.

Policy mixes

Policy mixes are comprehensive policy packages that effectively combine direct regulations, economic techniques, and other policy processes. The merits and drawbacks of the individual environmental policy measures that have been explained up until now are summarized in Table 4.1.1.

In recent years, as environmental problems grow worse, it is more important than ever to reflect on policy efficacy issues of the past, host careful deliberation regarding the unique nature of policy challenges at hand, and implement effective policies that allow the strengths and weaknesses of numerous processes to effectively compliment

Table 4.1.1 Policy advantages and disadvantages

Policy type	Advantages	Disadvantages
Economic methods	• Costs could go down • Stimulate technological innovation • Not much information required	• High administrative and oversight costs • Many political disputes and strife • In tax scenarios, difficult to predict society's total emissions
Direct regulations	• Enforceable by law • Easy to predict • Clear	• Large amount of information required • Does not stimulate technological innovation • High administrative and oversight costs
Voluntary regulations	• Cost effective • Peer organizations can apply pressure • Can integrate environmental issues into business processes	• Standards go down • Enforceability is weak • Transparency issues/responsibility to explain could become problematic

and reinforce each other. By doing this, policy makers aim to capitalize on the synergy produced from conjunctive regulatory frameworks.

From large-scale, long-term issues such as climate change, to other more localized air, water, and soil pollution issues, there is a very slim chance that companies and other organizations will adopt singular or unified economic approaches to addressing the majority of environmental issues. Indeed, policy measures that lump together direct regulations and other mechanisms are often more viable and make way for joint implementation alongside various economic policy measures. When multiple measures are adopted simultaneously, the strengths and weaknesses of each individual measure complement each other, increasing their collective efficacy.

Of course, policy mixes are not recommended when the various regulations conflict with one another or the detriments that they yield outweigh the merits. In other words, policies included in policy mixes must aim to make included policies bolster each other, make up for comparative drawbacks, and capitalize on relative merits. If this can be executed smoothly, even if the policy fails, the damages that arise will not be as great as those that would result from the failures of a single failed policy. Accordingly, it is imperative to form an inter-organizational network of governments, businesses, and other pertinent institutions that can draft and ensure compliance with reasonable regulations to be included within effective policy mixes.

Take waste and recycling policies, for example. A variety of mechanisms are often jointly employed to achieve overall policy aims, from direct regulations for preventing illegal dumping, to economic methods that levy processing fees, to procedural techniques for environmental management systems. This is clearly the case with the *Climate Change Tax* enacted in April, 2001 by the U.K. government. Its core focus is on environmental taxes on manufacturing, business, and public utility sector energy use (excluding the household sector). In addition, however, the taxes were implemented within the *Climate Change Protocol* policy package that includes a CO_2 emissions trading system. Aiming to compel businesses into compliance via emissions trading, the framework also provides an opportunity for companies that comply with policy mandates and other voluntary regulations to reduce their climate change by 80%. Of course, when companies are unable to adopt certain aspects of the protocol, they become ineligible for the reduced tax scheme for the subsequent two years. An added benefit to the system is that tax revenues help to finance the U.K. government's investment in renewable energy development and energy efficiency improvements.

SUMMARY

Issues that arise among reasonable environmental policy prescriptions are fundamental to the study of environmental economics. More than direct or technological regulations, economic measures are capable of accomplishing environmental policy targets in a cost efficient manner. However, when there is uncertainty about the marginal benefit, the comparative slope among the marginal benefit and marginal externality cost curves is what determines whether environmental taxes or emissions trading systems are superior. Moreover, when there is uncertainty about the marginal externality cost, neither environmental taxes nor emissions trading systems are necessarily preferable.

REVIEW PROBLEMS

1. What are the benefits and drawbacks of direct regulations?
2. Explain whether environmental taxes or emissions trading systems are ideal.
3. Why are policy mixes needed?

LEARNING POINT: WHY EMPLOY DIRECT REGULATIONS?

For a long time now, many countries have employed direct regulations as environmental policies. This is because they are easy to understand by society at large, and their goals are transparent. Conversely, environmental taxes have been applied to issues with little political resistance. Generally speaking, industries and citizens often view environmental taxes as aiming primarily to collect tax revenues and as ineffective for environmental targets.

Companies that earn profits or suffer large losses due to the presence of environmental policies have specific choice preferences among individual policy procedures. Those that earn large profits support such policies, but those that suffer losses aim to block their implementation. In short, organizations act in accordance with what is advantageous to them, working to promote, implement, repeal, or block what is disadvantageous. This is known as *rent seeking*.

For polluting firms, even when emission reduction targets are the same across policy choices, they still prefer certain policies to others. Specifically, they believe that subsidy systems that offer financial rewards are most preferable. After subsidies, they prefer direct regulations, and they prefer environmental taxes the very least. Why are direct regulations more appealing than environmental taxes? Well, since subsidies help to fund technological development, and direct regulations set technology standards, neither of them shut existing companies out of the policy deliberation process. Under such circumstances, firms can push for the standardization of technologies with which they possess comparative advantage, making it more difficult for new firms to enter the market. Thus, if the primary aim of existing companies is to preserve their privileges comparative to others, then direct regulation systems are preferred over environmental taxes. In this way, if government policy makers include existing firms in the policy deliberation process, they acquire a reputation of not aiming to implement policies that are best for society, but to implement those that will increase the profits of existing firms.

In subsequent policy studies, it is made clear that adopting economic techniques, more than direct regulations, achieves greater pollution reductions with fewer costs. For example, according to survey results comprising over 40 research projects performed by the U.S. Environmental Protection Agency (EPA), firm cost reductions amounted to 100 billion yen per year, and transaction costs were reduced by over 1 million yen. Furthermore, environmental problems are said to arise in the absence of markets for pollution management. Creating markets that incorporate pollution, and not direct regulations, incentivizes profit-maximizing frameworks and leads to efficient economies that utilize the environment in an optimal manner.

Section 4.2: Waste policy

Waste-processing system

Fees involved in waste-processing

Important waste issues, from decreasing the amount of waste disposed, to recycling, to the promotion of multiple-use of materials, are increasingly pertinent topics with regards to the creation of a sustainable society. The Japanese Ministry of the Environment stresses the importance of including adequate waste payment systems as cornerstones of the transition to sustainable society, thereby reducing waste and alleviating waste-related environmental harm. The fees levied for home and business waste-processing services vary with each municipality (city, town, village, etc.). Recent years are marked by a substantially greater uptake of waste-processing fee systems by local governments. However, while some regions have achieved wasted reductions of 10% over five-year periods via waste payment structures, there still exist many others that have experienced no change in waste levels since their structures were put into place. Said systems have thus acquired variegated reputations. By what standards should optimal waste amounts be determined in order to maximize social profits? Are garbage processing fee systems valid for checking waste disposal levels and maximizing social gains? In addition to deposit systems for harmful waste substances, what are some ideal fee systems and waste policies? In this section, we consider the answers to each of these prompts.

When waste-processing is free

There are facility, rent, fuel, and various other costs involved in waste collection and processing.

To date, throughout many of the regions performing waste-processing, local governments are deemed responsible for paying these costs by allocating residence and other tax funds. To address this concept, first consider a scenario in which *waste-processing fees* are not in effect.

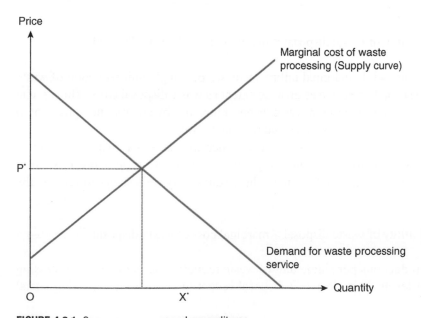

FIGURE 4.2.1 Company revenues and expenditures

In such scenarios, where the demand for waste-processing services is pegged to the amount of waste generated, the normal demand function generates a curve that tends downward and to the right (as depicted in Figure 4.2.1). Figure 4.2.1 includes a horizontal axis that measures waste amounts, a vertical axis that tracks the marginal unit price of waste, and a marginal cost curve for local waste-processing. Since waste-processing in this scenario is free, there is no incentive to decrease waste and household waste to X_M levels. Accordingly, locally processed waste levels reach X_M. Here, total social gains equal the area under the demand curve from the origin to waste level X_M. However, when waste levels amount to X_M, the social gains are not maximized. This is due to the fact that the costs to process waste allows for greater social gains in periods with less waste. In other words, when waste-processing is free, waste levels expand by the greatest amounts feasible, and there are excessive costs involved in waste-processing.

When there are waste-processing costs

To what extent should waste be reduced in order to maximize social gains? Consider a "pay-as-you-go" fee collection system, in which one is charged a certain amount based on the amount of waste one produces. The marginal cost curve of waste-processing proceeds up and to the right.

When waste disposal levels reach X_M, the marginal cost for waste-processing exceeds the waste-processing service demand. Since the demand curve stands in for the marginal utility curve, the marginal utility of waste-processing service demand among household budgets that is accrued from marginal waste disposal reductions drives waste-processing prices down. Such a scenario reflects the function:

marginal utility of waste disposal < marginal cost of waste disposal

Beyond this point, while marginal utility decreases per single unit reduction of waste disposal, it does not decrease to as great an extent as waste disposal costs. Thus, social gains per single unit reduction in waste disposal increase by an amount equivalent to the difference between marginal cost and marginal utility.

Now, is it also the case that social gains increase as waste disposal is cut to levels below X^*? In this case, the marginal cost of waste-processing is smaller than the demand for waste-processing. Thus, conversely, this scenario reflects the function:

marginal utility of waste disposal > marginal cost of waste disposal

Since utility reductions per single unit of waste reduction surpass waste-processing fee reduction levels, it is clear that disposal reductions lead to diminishing social gains.

In line with the above explanation, social gains are maximized when waste disposal levels are at X^*. In other words, when:

marginal utility of waste disposal = marginal cost of waste disposal,

waste levels maximize social gains. There, when processing fees per unit of waste disposal are set at P^* levels, it begins to cost households P^* per unit of waste, and as waste disposal levels surpass X^* levels, the amount of additional utility gained drops below P^* levels, which discourages further waste. This attests to the possibility of restricting waste disposal to optimal amounts while maximizing social gains.

However, nowadays, many localities levy fixed-price fees in which the waste-processing fee per household is collected on a monthly basis and does not change. In this scenario, even though payment structures are in place, since the cost for waste-processing does not increase hand in hand with greater amounts of waste, the marginal cost curve becomes horizontal, meaning its slope is always zero. Waste levels then reach X_M in the same manner as when waste-processing is free. This means that fixed-price fee systems are capable neither of checking waste levels nor maximizing social gains. Thus, even if they are implemented in a given municipality, the fact that they levy funds to pay for waste-processing does not make up for their inability to reduce waste amounts.

The main point is that, to the extent that processing fees for waste disposal are not burdensome, they fail to incentivize households to check their waste disposal. As a result, greater waste amounts lead to maximum use of landfill space, detracting availability for future generations. This, in turn, forces future generations into an over-reliance on recycling, which, regardless of feasibility issues, would prove to be extremely costly.

What are deposit systems?

Deposit systems in which initially collected fees are returned to consumers after proper disposal, are effective for guaranteeing that waste disposal/recycling organizations receive funds necessary to overcome substantial management costs. The aim of such systems is to collect and reuse resources in a manner similar to Figure 4.2.3, so a fixed fee (otherwise known as a deposit fee) is added to the sales prices, and the consumer, returns containers or materials (instead of simply throwing them away) and gets funds equivalent to the original fee returned to them.

Deposit systems lead to increased recycling and used material collection and processing rates, which is particularly effective in preventing unorganized mixing of post-consumption materials, ensuring that hazardous materials are not disposed with general garbage, and placing a check on littering.

Many deposit systems throughout Japan and the U.S. handle glass bottles and other beverage containers. Despite a lack of strong legal conventions, local shops often recognized the value of reusing glass and containers, propelling local and

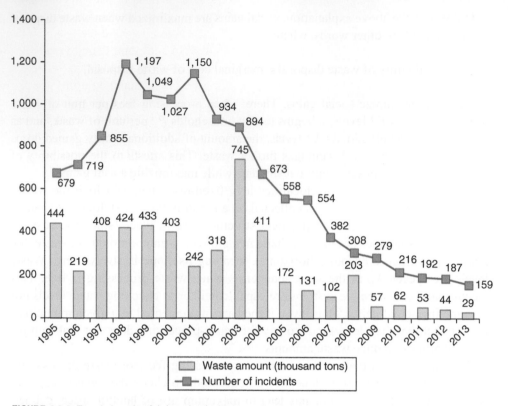

FIGURE 4.2.2 Trends in unlawful dumping incidents and waste amounts

Source: Ministry of Environment data

Note 1: Incidents where more than 10 tons of industrial waste was dumped were recorded by municipal and local health organizations. These data are included in the total number of dumping cases and waste amounts (however, data for cases where less than 10 tons of waste was dumped were also included for incidents involving special procedures for industrial waste processing).

Note 2: Over 16,000 tons of illegally dumped waste in Aomori and Iwate Prefectures is allotted to the year 1999. Another confirmed 860,000 tons dumped between the years 2002 and 2003 remain unaccounted for. Furthermore, 510,000 tons of waste dumped in Toshima, Kagawa Prefecture in the year 1990 has been confirmed yet exists outside of the scope of the above figure.

regional efforts to establish deposit systems instead of allowing said materials to be disposed conventionally. However, as glass bottles are relatively cheap, they tend not to be viewed as valuable resources. Even still, as the types of containers diversify, collection costs are viewed as a problem. And since the advent of single-use, non-reusable containers in the 1970s, returnable containers have experienced decreasing popularity popular as the deposit systems that ensure collection costs have often been abandoned.

To this day, there are many who demand that containers, dry-cell batteries, and other materials be properly recycled, yet these voices go unheard when makers and

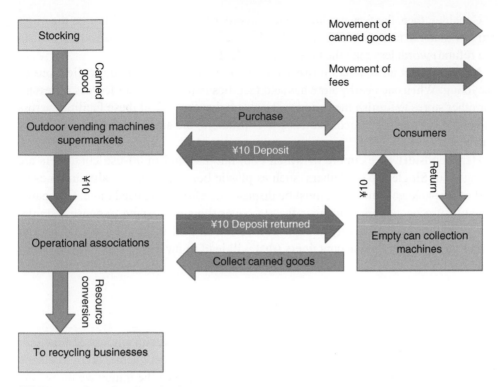

FIGURE 4.2.3 Deposit system structure

retailers object. However, many Western countries have successfully reduced waste levels through deposit systems for cans, glass bottles, and plastic bottles.

In theory, deposit systems could include both taxes (deposit fees) for such materials as canned beverages, as well as subsidies (deposit fee refunds) for the return of such used goods as empty cans. They could thus be a combination of Pigovian taxes and subsidies. To address the issue of littering, Pigovian taxes place fees on goods when they are purchased. Furthermore, subsidies are also effective for preventing littering by returning initially paid fees to consumers after collection.

Setting Pigovian taxes and subsidies to the same price is not necessarily required. This is because when containers or other goods are properly returned, there are collection, burial, or other processing fees involved while the subsidy (deposit fee refund) is only equal to the processing fee and ought to be less than the tax (deposit). However, if materials are recycled, the recycling price is roughly the same as the original raw material price, the processing price is counterbalanced, and Pigovian tax and subsidy amounts ought to be the same.

In the U.S. and Europe, deposit systems for cans, glass bottles, and plastic bottles have greatly encouraged recycling, leading to increased collection rates by 70% to 90% and largely curbing disorganized waste disposal. By contrast, the plastic bottle collection rate in Japan, where there is no deposit system in place, was about 17% as of 1998.

Consider, then, the fees producers incur in order to recycle. In addition to the original raw material price, when the price burden is exceptionally large, subsidies (deposit fee refund) worth less than the taxes (deposit fee) are applied.

Moreover, not all retailers possess the facilities needed to conduct collections or recycling. When one nearby store has said facilities in place, it becomes less necessary for other stores to furnish them, and to the extent that the lack of these facilities across the board is not a great inconvenience to consumers, the amount of money refunded can be cut.

Here, consider an example in which both one-use and multi-use containers are being sold. Single use containers, such as plastic bottles, steel and aluminum cans, and other package containers, must be disposed of after being used once. Compared to multi-use containers that incur fees to cover their processing expenses, single-use containers are relatively cheaper. Since this would inevitably lead to greater employment of single-use containers (that will lead to increased waste levels), they must be taxed.

Recycling policies

It is important to limit the number of landfills and allot a certain amount of final processing plant space for future generations. This can be largely achieved by reducing the amount of trash buried in the present. By recycling even the things we throw out, we can limit waste creation. Recycling is an alternative to disposal, and it is crucial to determine what kind of recycling accounts for costs and social benefits more efficiently than disposal. The most commonly recycled materials to date include plastic bottles, steel, paper and paper pulp, cloth (clothing items), cooking oil, aluminum cans, ink cartridges, and glass bottles (see Figure 1.2.3 for an overview of trends in plastic bottle recycling).

The benefit of recycling is that material items that would be disposed of are collected and reused, and as items that no longer need to be thrown out increase, the amount of waste incinerated and the amount of landfill space that is needed decreases, and overall disposal processing fees decrease. Other fees include the material collection fees and processing fees to prepare such materials for reuse. However, neither 100% nor 0% recycling rates are socially optimal. The aim should be to recycle at levels that maximize the social gains.

Chapter 1 provided an outline of optimal recycling levels. But at what point in the life of materials should recycling begin? Normally, cheaply procured raw materials (e.g. fossil fuels including natural gas) are exchangeable with recycled goods, so they share a competitive relationship. Consider oil, coal, and other non-renewable resources that have finite reserves and are consistently consumed. In the future, when said resources become scarce or they cannot be replaced, their scarcity value will cause them to be utilized in the most efficient manner. This has been depicted in the past by the explosive increase in oil prices that drove many renewable energy resource undertakings as substitutes to oil. Plastic and aluminum and other raw materials are subject to the same conditions of cause and effect.

In cases where recyclable materials are interchangeable with raw materials, the former are not recycled if they are not cheaper than the latter. Yet, procuring raw materials also yields pollution. By imposing taxes equivalent to externality costs, raw materials become more expensive, and recycling becomes comparatively cheaper for society. Finally, if raw materials are replaceable, the relative price of replacement goods determines whether or not they are indeed supplanted.

Of course, when raw materials are abundant, their prices are low. And furthermore, some raw materials have historically been given preferential treatment through favorable subsidy and tax policies. For example, lumber industries in many developed countries receive subsidies for their manufacturing activities. This leads to a downward shift in the supply curve of paper. Thus, the price of paper decreases and its production increases. In many cases, there is also diminished, if any, incentive to use recycled paper.

In order to promote recycling, preferential treatment subsidy and tax policies that minimize raw material prices must be abolished. Pigovian taxes on raw materials would also serve to encourage recycling. In general, economic policies for promoting recycling include imposing collection fees to curb waste disposal and taxing raw materials to discourage the use of raw materials.

Greater numbers of municipalities are charging fees for waste-processing. However, waste fee systems are not without their problems. For one, there are many systems that only impose fees on certain waste amounts, which this section has proved to be the mark of sub-optimal, ineffective policy. This is especially true in many communities throughout Japan. For, although the national average processing fee for one bag of garbage is around 400 yen, many municipalities only charge around 40–80 yen. This waste-processing fee is largely below optimal levels, so there are no incentives to reduce waste to appropriate levels.

When there are no processing fees for waste disposal in place, regional recycling initiatives that do not employ financial penalties could adopt tax exemption and subsidy support structures. However, simply setting the recycling rate does not become adequate incentive, leading to minimal recycling. Tax exemption for installing recycling installations, subsidies for buying equipment, and providing cheap land for constructing recycling facilities are various measures for promoting recycling processes. However, as these systems do not involve direct recycling, they run the risk of promoting recycling facilities that are too capital or land intensive, which would not necessarily be optimal for recycling businesses themselves.

SUMMARY

When there are no processing fees applied to waste disposal, there are no incentives for households to limit the amount of waste that they produce. Thus, fee systems are key to curbing disposal. Moreover, deposit systems are effective for material collection and reuse schemes.

REVIEW PROBLEMS

1. Explain the benefits of the measured rate system vs. the fixed rate system for garbage fees.
2. What penal regulations serve to check unlawful littering?
3. Within which conditions are deposit systems ideal?

LEARNING POINT: PENALTIES FOR ILLEGAL DUMPING

Evading garbage processing fees to reduce one's own garbage expenses in cases where there are appropriate costs attached to processing, and cases where people feel as if disposal is inconvenient because dumping grounds are so far away are both examples of circumstances that give rise to *illegal dumping*. Illegal dumping takes on many forms, from littering, to construction waste dumping, to throwing away major household appliances (air conditioners, cathode-ray televisions, electric refrigerators and electric freezers, and electric washing machines).

Japan enacted a household appliance recycling law (*Law for Recycling of Specified Kinds of Home Appliances*) on April 1st, 2001. It promotes the recycling

Table 4.2.1 Electrical appliance recycling and disposal fees

Target product	Collection/Transportation fee	Recycling fee	Cost to the consumer
Televisions (cathode-ray)	¥3,150 (tax included) for models 21 inches and below ¥5,250 (tax included) for models 24 inches and above	¥2,700 to ¥3,615	¥5,850 to ¥8,865
Refrigerators	¥3,500 or more	¥4,600 to ¥5,590	¥8,100 to ¥9,090
Washing machines	¥3,500 or more	¥2,400 to ¥3,280	¥5,900 to ¥6,780
Air conditioners	¥3,500 or more	¥3,500 to ¥4,490	¥7,000 to ¥7,990
Personal computers		¥3,000 to ¥4,000	¥3,000 to ¥4,000

Source: Produced by the author with values calculated by producers and associations

of the components and materials of used household appliances thrown out at homes and offices, and aims to reduce waste amounts and to utilize resources more efficiently. It also makes those who dispose of household appliances to pay fees (as depicted in the household appliance disposal/recycling fee examples in Table 4.2.1).

There are two types of externality costs that arise from illegal dumping. The first is that someone other than the dumper is burdened with the processing cost that ought to have previously been accounted for. The second is the worsening of the environment, which, beyond the health risks posed, includes people's inability to bear the worsening physical appearance of the surroundings. It is essential that firms and governments adopt policies that correct environmental degradation caused by illegal dumping by enforcing monetary penalties.

Illegal dumping policies could indeed impose fee penalties. In addition to continuing fee systems that are already in place for waste disposal, it is essential that the monetary punishment for illegal dumping is sufficiently large. The reason why is if the anticipated cost of illegal dumping that is calculated by multiplying the probability of being caught by the fine imposed is not larger than the base disposal fee, then there is a large chance that illegal dumping will occur.

There are many large penalty fee systems in place across numerous countries and regions. For example, with Japan's *Waste Management and Public Cleaning Law*, illegal dumping fines of up to 100 million yen can be calculated. Furthermore, increased stress placed on strengthening supervision in order to discover illegal dumping has made the probability of being caught greater. However, there have actually only been a few cases where fines have been imposed. In the case of Singapore, a first offence of littering costs 1,000 Singaporean dollars, while a second case costs 2,000 Singaporean dollars. Thus, Singapore has largely prevented littering issues.

Section 4.3: The Kyoto Protocol and climate change policies

The climate change issue

As noted in Chapter 1, climate change remains an unresolved issue. The global environment is non-excludable and non-competitive by nature, and impacts on it transcend national borders, rendering it a worldwide public good. When global environmental preservation policy is lacking, and voluntary action by individuals and enterprises is relied upon, precession for society is at minimal levels. Because of that, it is necessary to implement some sort of public policy. Currently, the policy that addresses the climate change issue is being debated across the globe.

In 2005, the Kyoto Protocol (the formal name for it being the Kyoto Protocol to the United Nations Framework Convention on Climate Change) took effect and with it began actual movements to decrease greenhouse gases. Kyoto Protocol refers to a protocol adopted at the Third Session of the Parties to the United Nations Framework Convention on Climate Change (Kyoto Conference on Climate Change, COP3). With the Kyoto Protocol, numerical reduction targets for the greenhouse gases that cause global warming (CO_2, methane, and others) and methods for achieving such targets were specifically set. The numerical targets for developed countries were a 5% decrease in greenhouse gas emissions compared to 1990 levels. As displayed in Table 4.3.1, the reduction targets for each country were set, and in Japan, the Kyoto Protocol objective implementation plan was made. Moreover, the emissions reductions targets for developing countries like China were not set by the protocol.

The Kyoto Protocol and Kyoto mechanisms

The Kyoto Protocol

The Kyoto Protocol came into existence after the adoption of the United Nations Framework Convention on Climate Change that took place at the Earth Summit in Rio de Janeiro, Brazil, in June of 1992. There, countries from across the globe confirmed methods to make stable yet progressive efforts to reduce the concentration of greenhouse gases in the atmosphere. Upon the COP3 in December of 1997, the Kyoto Protocol was adopted. While the protocol was adopted, it was not immediately implemented due to abstentions from the U.S. and Russia. After some time, the Protocol was officially implemented on February 16, 2005, 90 days after it was ratified by Russia on November 18, 2004.

The Kyoto Protocol was critically important as one of the first arrangements to build international consensus in setting specific greenhouse gas reduction targets. And, as a precedent-setting first step of climate change policy, it helped to unify the international community in its considerations of relevant geographical information, value judgments, benefits and losses, and so on. Furthermore, it transparently forces countries to absorb the costs of greenhouse gases, and regardless of the presence of large uncertainties surrounding climate change, there was inherent value to ratifying it.

The Kyoto Protocol is widely known for its flexible economic policies known as Kyoto mechanisms. These are policy measures that make use of market mechanisms

Table 4.3.1 Kyoto Protocol emissions reduction targets

Country	Numerical target	1990 GHG emission level
France	0.0%	568.0
U.K.	−12.5%	748.0
Germany	−21.0%	1,243.7
EU total	−8.0%	4,240.0
Russia	0%	3,046.6
Japan	−6%	1,187.2
U.S.	−7%	6,082.5
Canada	−6%	595.9

Source: Kyoto Protocol

1-3 Signatory I Country List

Numerical targets of greenhouse gas emissions (compared to standard year emission levels) among Signatory I Countries are listed below.

→ Through the Kyoto Protocol, the numerical target for 15 EU nations was −8%, but the numerical targets of each individual country were reanalyzed (and recognized by the Kyoto Protocol) and published here.

→ Countries highlighted in blue had not ratified the Kyoto Protocol as of 2006

→ The source for each country's 1990 GHG emissions level (units: 10,000 tons of CO_2) is <FCCC/SBI/2005/17>, and these numbers differ from the standard year emission level values

→ Countries shifting to market economies where the standards for CO_2 emissions levels exists outside of the year 1990 are Bulgaria (1988), Hungary (1985-87 average), the Netherlands (1988), Romania (1989), Slovenia (1986)

→ For Croatia, Romania, Liechtenstein, and Monaco, there are reduction targets as Annex B countries of the Kyoto Protocol, which are not included in the climate change frameworks for signatory I countries

→ There are no numerical targets for greenhouse gas emissions for non-signatory I countries

→ There are 128 countries ratifying signatory I countries of the Kyoto Protocol as of 7/10/2006

EU countries (15 Kyoto Protocol signatories)			Countries shifting to market economies (EIT)			Other countries		
Country	Numerical target	1990 GHG emission level	Country	Numerical target	1990 GHG emission level	Country	Numerical target	1990 GHG emission level
Portugal	27.0%	53.9	Russia	0%	3,046.6	Iceland	10%	3.3
417	25.0%	109.4	Ukraine	0%	978.9	Australia	8%	417.9
Spain	15.0%	283.9	Croatia	−5%	31.8	Norway	1%	50.1
Ireland	13.0%	53.8	Poland	−6%	564.4	New Zealand	0%	61.5
Sweden	4.0%	72.2	Romania	−8%	265.1	Canada	−6%	595.9
Finland	0.0%	70.4	Czech Republic	−8%	192.0	Japan	−6%	1,187.2
France	0.0%	568.0	Bulgaria	−8%	138.4	U.S.A.	−7%	6,082.5
The Netherlands	−6.0%	211.7	Hungary	−6%	122.2	Switzerland	−8%	52.4
Italy	−6.5%	511.2	Slovakia	−8%	72.1	Liechtenstein	−8%	0.3
Belgium	−7.5%	145.7	Lithuania	−8%	50.9	Monaco	−8%	0.1
England	−12.5%	748.0	Estonia	−8%	43.5	Turkey	−8%	
Austria	−13.0%	78.6	Latvia	−8%	25.4			
Denmark	−21.0%	70.7	Slovenia	−8%	20.2			
Germany	−21.0%	1,243.7	Belarus		129.2			
Luxemburg	−28.0%	13.4						
EU total	−8.0%	4,240.0						

in order to allow countries to achieve their individual numerical targets. Figure 4.3.1 outlines the three Kyoto mechanisms: *Joint Implementation (JI)*, *Clean Development Mechanisms (CDM)*, and *emissions trading systems*. With the Kyoto Protocol, climate change negotiations set the numerical targets greenhouse gas emission reductions for developed countries (Annex I Parties). However, in countries like Japan, high energy efficiency had already been achieved due to the development of energy conservation technologies, so achieving such numerical targets only through domestic endeavors posed a serious challenge. Furthermore, countries with abundant room for efficiency improvements (such as China) could take on and achieve such targets much more cheaply, so investing in the realization of greenhouse gas emissions in other countries became recognized as a valid endeavor.

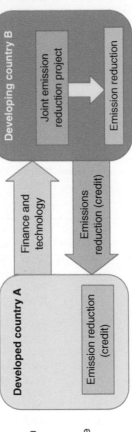
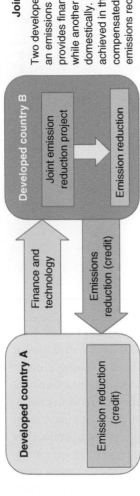
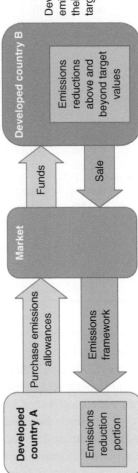

Joint implementation (JI)

Two developed countries cooperatively execute an emissions reduction project. One country provides financial and technological support while another implements the project domestically. When emissions reductions are achieved in the latter, the former are compensated for their contributions with emissions reduction credits.

Developed country A

Emission reduction (credit)

Finance and technology

Emissions reduction (credit)

Developed country B

Joint emission reduction project

Emission reduction

Clean development mechanisms (CDM)

One developed and one developing country cooperate to execute an emissions reduction project. The developed country provides financial and technological support while the developing country implements the project domestically. When emissions reductions are achieved in the latter, the former are compensated for their contributions with emissions reduction credits.

Developed country A

Emission reduction (credit)

Finance and technology

Emissions reduction (credit)

Developing country B

Joint emission reduction project

Emission reduction

Emissions trading

Developed countries buy and sell emissions levels in order to reach their individual emissions reduction targets.

Developed country A

Emissions reduction portion

Purchase emissions allowances

Emissions framework

Market

Funds

Sale

Developed country B

Emissions reductions above and beyond target values

FIGURE 4.3.1 Kyoto mechanisms

Source: METI, Environment Section of the Industrial Structure Council, 2004 (http://www.meti.co.jp/committee/downloadfiles/g41104a91j.pdf)

Kyoto mechanisms

Kyoto mechanisms are aimed at advancing climate change policy objectives by reducing overall costs to the lowest possible levels through projects that enhance technology, information, and capital sharing among multiple involved countries.

Joint implementation

Joint implementation (JI) is a system through which two developed countries (including those in the Central and Eastern Europe, Russia, and other places that have made the transition to market economies) jointly execute greenhouse gas reduction or absorption projects. This is mainly accomplished as one foreign investing country earns credit for the emission reductions achieved in the partner country in which they invest. For example, when Japanese industries invested in emissions reductions projects in Russia, a portion of the reduction amounts were recognized as having been made by Japanese businesses. Specifically speaking, countries that contribute to projects abroad are called investor nations, while those that implement projects domestically are called host nations. Moreover, both nations involved in JI projects are developed countries. However, in such scenarios, since the emissions framework and numerical targets that are set among developed countries are acquired and transferred, the overall amounts of emissions frameworks do not change.

Clean development mechanisms

Clean development mechanisms (CDM) are systems through which developed countries that have numerical targets for greenhouse gas emission set by the Kyoto Protocol invest in greenhouse gas emissions reduction projects in developing countries that have no emissions regulations. Then, the resultant emissions reductions (subtracted from the regulatory baseline; the baseline is the anticipated greenhouse gas emission conditions that would arise should the CDM project not be implemented) achieved within developing countries can be earned as reductions credits by the developed countries that invested in the projects. The aim of these mechanisms is to contribute to global warming policy and induce sustainable development among developing countries.

Through CDM, developing countries that have high marginal emission reduction costs for greenhouse gases can uphold their emissions regulations in a less costly manner. They allow for developing countries to receive technological and financial support from developed countries, tying them to emissions regulations, and contributing to overall global emissions reductions.

At a glance, the results of CDM would make one believe that they are systems with inherently large benefits. However, some of the problems with CDM include political issues that arise from CDM implementation conditions and regulatory baseline setting. Furthermore, the majority of the most recent projects chemically process and convert greenhouse gases with larger global warming impacts than CO_2 (such as methane and furan) into equivalent amounts of CO_2. If these pre-processed gas amounts were left untouched, their single unit impact on global warming would be greater than that of CO_2. Yet converting these gases to CO_2 equivalent amounts result in reduction credit awards. For example, the CDM credit (rights to reduction and absorption amounts) for hydro fluorocarbons (HFCs) and nitrous oxide (N_2O) reductions alone make up over 70% of total reductions (see Figure 4.3.2).

One of the political reasons for adopting CDM as a Kyoto mechanism is that they are aimed at resolving friction about global warming policy between developing and developed countries. Modern developed countries emitted the majority of the CO_2 and other greenhouse gas amounts that have accumulated in the atmosphere in the past. Moreover, much of the emissions amounts that continue to this day are from developed countries.

Even still, the reason that developing countries object is that the damages dealt by climate change will only appear in the mid to long term, 50 to 100 years from now. Thus, they often do not see any value in paying the present costs of adhering to related policies. The future benefit of avoiding the effects of climate change many years down the road is an exchangeable ratio of present to future resources for society as a whole. It is evaluated according to the present discount value that is discounted by the social discount rate. However, while the future benefit of climate change prevention is the

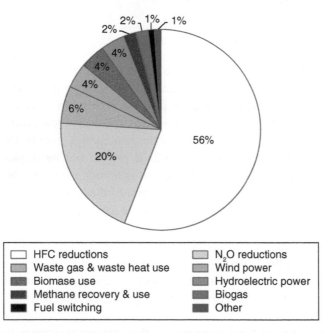

Based on projects that have already been executed or are at some stage of implementation: 744 cases, 142 million CER/year.

FIGURE 4.3.2 Breakdown of CDM project fields

Source: CDM Board of Directors Documents

same for both developed and developing countries, due to discount rates, the present discount value of developing countries is less than that of developed countries. The social discount rate is dependent upon a country's marginal productivity of capital and time preferences. In developing countries, compared to other resources (such as labor), capital is less abundant, so the marginal productivity of capital (or the earning rate of capital) is always increasing. Therefore, the discount rate in developing countries becomes high, and the discounted present value becomes smaller than that of developed countries with low marginal productivity of capital. In the end, developing countries prioritize their own economic growth, and there become few incentives to implement climate change policies.

Still, there is high potential for developing countries to check emissions through CDM because it assists them in establishing finance and technology-based partnerships, leading to incentives to ratify the Kyoto Protocol. In order to promote further participation in future climate change policy, participation among developing countries will be indispensable. From this perspective, there exist substantial political advantages to backing CDM.

Emissions trading

Finally, there is emissions trading. Through emissions trading systems, portions of emissions frameworks can be bought and sold among developing countries that have standardized the emissions frameworks (for details, refer to Chapter 3, Section 3.3).

The significance and limitations of the Kyoto Protocol

One of the benefits of Kyoto mechanisms is that developed countries may be able to achieve their domestic emission reduction obligations at comparatively lower cost. The effects of JI, which essentially minimize emissions curtailment costs, also aid in preventing domestic enterprises from shifting their production centers to developing countries and lessens the potential for "leakage." Furthermore, CDM, as a unique and flexible system that incentivizes climate change policies in developing countries while minimizing costs for developed countries, show signs of garnering further international support in the future. These benefits, collectively, lend credit to Kyoto Protocol implementation.

However, in order to prevent the direst effects of climate change, the atmospheric CO_2 concentration would need to be pegged to the 280 ppm levels present in the latter half of the 18th century. Realizing this would require the international community to cut current total global emissions amounts in half. Moreover, the IPCC (Intergovernmental Panel on Climate Change) reports that if global emissions levels were to continue at 1994 levels, the atmospheric CO_2 concentration would reach 500 ppm. These facts in conjunction with the fact that there are no determined emissions targets for developing countries, serve to undermine the overall efficacy of the Kyoto Protocol in curbing climate change. With an understanding of the Protocol's shortcomings, it is clear that in future, supplementary policy action is required. Accordingly, the international community continues to share information and debate the *post-Kyoto Protocol* policies that ought to succeed the Kyoto framework.

The Kyoto Protocol and environmental economics

The significance of the Kyoto Protocol when viewed from the lens of environmental economics is that it surpasses the single-country framework inherent in the emissions trading systems that have been proposed and experimented with since the 1970s. In effect, the Protocol makes said systems more practical, and by making use of market mechanisms, it is capable of measuring issue resolutions.

Consider the EU-ETS (European Union Emissions Trading Scheme) precedent. The EU-ETS began in January of 2005, and 27 European Union countries are included in its scope. Phase 1 took place from 2005 until 2007, while Phase 2 took place from 2008 until 2012. Member nations set domestic plans and distributed permissible emissions amounts to enterprises in accordance with the EU-ETS. Phase 1 targeted CO_2 from large-scale emitters in manufacturing industries that consume vast quantities of electricity, heat, and primary energy. This included fuel plants, petroleum coke purifiers, steel working factories, and cement, glass, brick, ceramic, paper pulp enterprises. Even though the scope was limited in this manner, it included over 12,000 manufacturing facilities across the entire E.U., constituting nearly 45% of Europe's total CO_2 emissions.

Phase 2 allowed for emissions transactions between non-governmental organizations, individuals, and enterprises by expanding the scope of participatory industries via revisions in 2006. Moreover, third-party countries anticipated being involved in emissions trading systems through JI and CDM. In fact, consultation with Norway and other countries involved in these frameworks still persists today. The results of the EU-ETS were:

1. Emissions reductions costs decreased,
2. Technological innovation was incentivized among participatory enterprises, and
3. The price of CO_2 was more widely recognized as prices were affixed to emissions rights.

The EU-ETS was recognized as a more cost-effective method than the emissions trading framework of the European Council. To credit it further, due to its legacy and ability to set the stage, the EU-ETS allowed for Kyoto Protocol targets to be achieved with 2,900 to 3,700 million Euros/year as opposed to the 6,900 Euros/year that it would cost were it not in place. And finally, the system is hailed for having increased business environmental consciousness and incentivized them to be more proactive at reducing emissions and adopting cutting-edge technologies.

However, due to the recent economic downturn in Europe, production levels have dropped beyond what had been, leading to fewer production-based emissions and less demand for emissions rights through market transactions. In fact, many now

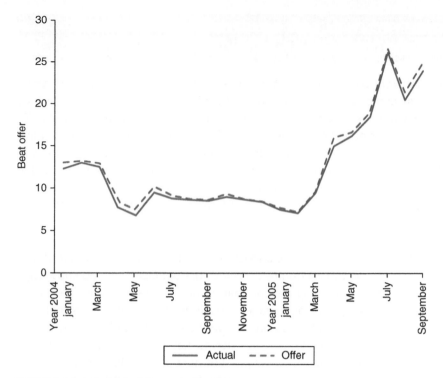

FIGURE 4.3.3 Emissions privilege price trends

Source: EU-ETS

forecast a sharp drop in emissions right transaction prices due to their existence in excess (see Figure 4.3.3).

SUMMARY

The global environment is non-rivalrous and non-excludable by nature, and its influence transcends national territories as a global public good. Currently, policies to confront the climate change issues are being debated in countries across the world. The Kyoto Protocol aimed to reduce emissions among developed nations by 5.2% of 1990-year levels. In order to achieve that target, new Kyoto mechanisms (joint implementation, clean development mechanisms, and emissions trading systems) that make use of market principles were implemented.

REVIEW PROBLEMS

1. Summarize the Kyoto Protocol and its contents.
2. What are Kyoto mechanisms?
3. Describe the problems with clean development mechanisms.

LEARNING POINT: CLEAN DEVELOPMENT MECHANISMS (CDMS)

CDMs have inherent issues. The first of these generally shows its face in emissions trading systems with market features. That is, CDMs are often plagued by information asymmetries that arise during emission credit measurements. To achieve sustainable development, even valid CDM project proposals must have their importance verified. Furthermore, there must be an assurance no organizations will receive special privileges. Compared with developing countries, these factors only dissuade developed countries and businesses from implementing CDM projects. Conversely, reports show that there is a greater incentive for firms to hide their emissions amounts in order to gain as many emissions credits as possible.

There is also the issue of power relationships within transactions. Since developed nations and firms in those countries possess greater CDM-related knowledge and monitoring ability, developing nations that commit to deals within these contexts do so at comparatively disadvantaged positions. The transactions that take place under these circumstances thus present various, unavoidable problems. The inspections, information gathering, and monitoring needed to solve these issues cost money. These costs are added to the procurement cost of CDM as transaction costs. These transaction costs can diminish with experience, but they need to be reduced as by much as possible at the earliest possible stage.

The second major issue is policy redundancy. This involves the proper means for separating projects that could be accomplished outside of CDM frameworks from those that would not be feasible without CDM. Moreover, projects that are highly feasible outside of the realm of CDM need not be adopted by climate change policies. And, although on-the-ground CDM operations are determined by governing boards that typically meet once every two months, their mandates to provide evidence of redundancy could produce lengthy discourse through the future.

Environmental project redundancy in which environmental impacts may be achieved with or without CDM frameworks, investment project redundancy where profits can be expected from investments made from within or outside of CDM frameworks, and financial project redundancy depicted by the existing official development assistance (ODA) are all examples of redundant projects that are not tied to CDM for their execution. The evaluation of these types of projects lies in whether or not they can be implemented as effective climate change policies. In cases where the redundancy checks are not in place, considerable credits arise among projects that do not impact climate change, and social benefits decrease.

However, countries like India that want to implement large amounts of CDM oppose resolution policies for these problems, contributing to a number of difficult-to-resolve political challenges. By ignoring the issues with redundancy, however, greater numbers of low-quality (contributions to climate change policy and sustainable development) credits emerge within the total system credit supply, resulting in declining credit values and losses for all developing countries.

Section 4.4: The current status and future challenges of climate change policies

Climate change policy initiatives in various countries

What are the most commonly applied climate changes policies, and are they constructed with Kyoto mechanisms in mind? These considerations are addressed in this section by shedding light on existing policy challenges. To lay the foundations for this discourse, this section opens with explanations of the political climate and policy initiatives in Japan, the E.U., the U.S., China, India, and Russia.

Japan

In accordance with the Kyoto Protocol, Japan has made efforts to reduce greenhouse gases by 6% of 1990 levels. However, these efforts have been relatively difficult compared to other countries. The first reason for this is that the marginal cost of CO_2 reductions for Japan is quite high. According to the IPPC report, in cases where CO_2 emissions were to be decreased through domestic measures alone, out of Japan, the E.U., and the U.S., the marginal abatement costs for Japan are the most expensive overall.

The second reason that emissions cuts have been difficult to achieve is that Japan does not have the same comparative advantages as Europe. The 1990 level targets are not appealing for Japan, and moreover, there doesn't exist a system that can make use of flexible global warming measures as they do in the E.U. bubble that shall be mentioned again later. While the E.U. has an advantage in the absence of U.S. participation, Japan faces the disadvantageous conditions, having to pay high costs and really endeavor to realize reductions. Nowadays, the emissions restrictions geared towards previous targets have been put on hold, and in 2010, emissions levels had only been decreased by 0.3% compared to 1990 levels. Because of this, the perfection of environmental taxes,

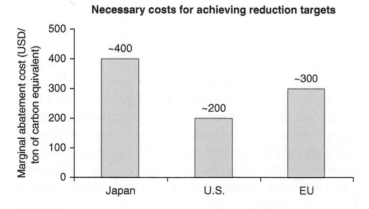

FIGURE 4.4.1 Regional comparison of marginal abatement costs

Source: Regarding National Countermeasures for the Global Warming Problem, Japanese Business Federation Outlook, September 19, 2001

emissions trading systems, and other domestic policies, as well as clean development mechanisms must be implemented via Kyoto mechanisms that are less costly.

The European Union

E.U. courts have been pushing for climate change policy since the early 1990s. They ended up successfully implementing a carbon tax, and more recently, the emissions trading system that it has set up (inducing trading between particular countries and throughout the E.U. as a whole) has become exemplary. Based upon the system configuration, compared to other countries and regions, it is rather progressive. And moreover, CDM is being proactively utilized. Because of technological exchanges, CDM oversight committees have recognized three out of four of top CDM projects as involving European nations, and the projects are continuing progressively (as of 2008). However, one must also consider how unfair it is that the E.U. is in a favorable position with regards to the Kyoto Protocol.

There are two reasons for this, and the first is the emissions standard years being set at 1990. As a result of fuel transitions (such as the shift from coal to gas), after U.K. emissions peaked in 1990, CO_2 levels began to drop.

The second reason is due to the existence of the so-called *E.U. bubble*. The E.U. bubble is a system under which emissions amounts across the entire E.U. are regulated. Because of this, even though there are countries that emit excessively compared to 1990 levels, the E.U. can uphold preserve emissions levels with other countries within the block that conversely have emissions amounts to spare (economic transition countries), so the result is that the E.U. is in a comparatively favorable position. Amidst this, the EU-ETS (European Union Emissions Trading System) was also implemented. From then on, there was the chance that the number of E.U. signatory countries would increase, and if that were happen, the E.U. bubble would become more flexible. If the unfairness of this is not improved, when international initiatives are stipulated, this could end up being a barrier to a certain degree (rebukes from or diminishing numbers of participatory countries, etc.).

United States of America

Emissions from the U.S.A. comprised 18% of world totals, making it the second largest emitter (see Figure 4.4.2). While the U.S. projected a pro-Kyoto Protocol stance in 1997, it eventually did not ratify because of the negative impacts it would have on the domestic U.S. economy and the lack of regulations for developing countries. However, due to public opinion shifts since Hurricane Katrina in 2006 and the sharp energy prices that followed, there have been ample opportunities to push for climate change policy and bolster natural disaster resilience. For instance, seven northeastern states agreed to establish an independent emissions trading system, which attests to growing public concern for climate change.

Furthermore, the U.S. Federal Government announced plans to increase economic growth and technological development through the Asia-Pacific Partnership for Clean

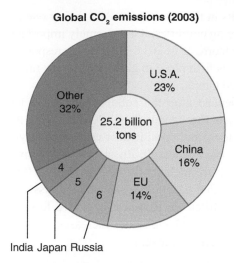

Global CO$_2$ emissions (2003)

U.S.A.
23%

Other
32%

25.2 billion
tons

China
16%

4

5

6

EU
14%

India Japan Russia

FIGURE 4.4.2 Breakdown of global CO$_2$ shares

Source: Produced from the Energy Economics Statistical Outlook, 2006 Edition

Development and Climate (APP). While this pact runs contrary to the Kyoto Protocol, it is nevertheless a considerable method for accelerating international initiatives to tackle climate change.

China and India

China and India are currently classified as developing countries, aiming for economic development and comprising a combined population of over 2,400,000,000 people. Incidentally, China now emits more CO$_2$ than any other single country in the world. The high costs and negative economic impacts that climate change policies would have on China and India, both of which are unwavering in their developmental aims, leaves little incentive for them to ratify climate change policies. However, in recent years, both countries have been quite responsive to invitations to CDM-based policy measures.

The number of CDM projects being implemented in both of these countries is increasing over time. All the while, their ranking among host countries demonstrating smooth and effective CDM projects is also improving. Moreover, based on their interest in APP, which would incentivize development through technological innovation, China and India have also shown increased concern for energy conservation and environmentally friendly technology. Yet overall, for either of these two nations to fully adopt climate change policies, developed countries must play a role in providing technological cooperation and economic incentives.

Russia

Compared to its massive territory, the scale of the Russian economy is relatively small, while its population scale is considerably large. This alludes to the fact that Russia, on the whole, will end up emitting more pollutants in the near future. And, being one of the world's foremost gas and petroleum producing countries, it is extremely important to constructing international policies. For the record, as of 2011, Russia was responsible for 5% of total global CO_2 emissions. That is fifth place in global emissions rankings, surpassed by India, the E.U., the U.S., and China.

Following the economic disorder it experienced after the collapse of the Soviet Union, Russia continued to curtail its emissions throughout the 1990s. In 1999, its CO_2 emissions amounted to about 1.8 billion tons, and considerable emissions privileges were sold to such other countries as Austria. By default, Russia was in a prime position to supply a great deal of emissions rights through the Kyoto Protocol. Be that as it may, Russia is often criticized for the fact that despite the sale of its emissions rights, it plays no role in reducing emissions. Moreover, Russia is expected to seek further economic growth, giving rise to concomitant CO_2 emissions increases, rendering it a key actor during Kyoto Protocol ratification discourse.

Difficulties with international cooperation

Establishing international initiatives like the Kyoto Protocol is extremely difficult. To date, the number of important players has increased; regardless of whether or not they participated in the Kyoto Protocol, every country considers its national interest as a pretext to executing global warming policy.

For example, while the U.S. refused to participate in the Kyoto Protocol because of the negative impact it would have on its economy, as energy prices continue to be high, energy conservation technologies have come to be recognized as critical. Furthermore, as China, India, and other developing countries continue to achieve economic development, there will be shortages in global energy, and there is a chance that this will translate into energy issues for the international community. Because of this, due to rising energy prices, proposals to develop and diffuse energy conservation technologies through technological agreements have been gaining popularity.

Through the Kyoto Protocol, the E.U. secures favorable conditions compared to other countries. Furthermore, through the sale of hot air, Russia can earn revenues. China and India can acquire technology and finance from developed countries through CDM. In this way, it will be extremely difficult in the future to create more realistic conditions for proceeding forth from the pretext that it is necessary to realize the individual interests of particular countries.

Issues with the Kyoto Protocol

Considering the difficulty of what every country thinks and stresses, the Kyoto Protocol was said to lead to many problems. There are various debates that focus on the problems with the Kyoto Protocol, but here, we will provide the main two.

The first problem is the validity of the Kyoto Protocol's global warming policy. The Kyoto Protocol creates the responsibility for developed nations to reduce overall emissions by 5.2% of 1990 levels by the year 2012. Yet as stated before, the U.S., as the world's largest CO_2 emitter at the time, refused to ratify. Considering these current conditions, regardless of how well Japan and the E.U. accomplish their targets, if the U.S., as the largest emitter, does not make reductions, and furthermore, if China and India increase emissions, then conversely, overall greenhouse gas emissions stand to increase.

The second problem is the lack of *participation incentives*. The Kyoto Protocol is legally binding, and if developed countries fail to comply with reductions commitments, they are subject to penalties. In addition to costly greenhouse gas reductions, there are little incentives to join a legally binding protocol. Thus, even to increase the number of participatory countries in the future, if it is legally binding, then the policy must put some sort of incentives in place.

Post-Kyoto Protocol options

The Kyoto Protocol includes a framework through the year 2012. Recently, throughout ongoing lively debate, there have been issues regarding what should be done in the "*post-Kyoto Protocol*" (also known as "post-Kyoto" or "post-2012") era after the year 2012. There are two main schools of thought regarding post-Kyoto choices. One is to extend and improve upon the current Kyoto Protocol, while another is to design a new system as a replacement for the Kyoto Protocol.

Extending and improving the Kyoto Protocol

The first choice being considered is to extend and improve the Kyoto Protocol. If a new policy were to be created from scratch, it would be politically complicated to set the next set of ordinances, and one could imagine that it would take time. It took three and a half years until the Kyoto Protocol was chosen, and from there, it took another three years for the Marrakesh Agreement (comprehensive rule-making of the Kyoto Protocol). Constructing a new framework would undoubtedly take a long time. If the Kyoto Protocol were extended, that time could be reduced.

However, extending the Kyoto Protocol brings with it great challenges to overcoming its shortfalls. The first of these challenges is the need to increase the number of signatory countries. Consider, for example, the U.S. and its position as a world leader. Its choice not to ratify the Kyoto Protocol severely undermined the potential and efficacy of the overall program. While participating countries collectively account for 55% of global emissions, further increasing the number of signatory countries is critical to achieving long-term emissions reduction targets. Specifically, heavily polluting developing countries, operating outside of the Kyoto framework in their developmental goals, ought to be encouraged to participate. It goes without saying that emissions targets for such rapidly improving countries as China and South Korea should be established.

The second option is to revise the Kyoto mechanisms. As previously stated, Kyoto mechanisms allow for flexible approaches to cutting the costs of CO_2 emissions reductions among developed countries. On the other hand, many problems exist in emissions trading and CDM schemes. Determining the baseline transaction costs and the validity of climate change policy are some of the challenges inherent in CDM. Solving these would allow for accelerated climate change policy approval.

Finally, there must be a long-term plan that garners incentives for each participating country. The Kyoto Protocol prioritized feasibility over effectiveness. In other words, while it did successfully provide a global framework, it often gave rise to weak climate change policies in order to guarantee widespread, multinational participation. Thus, while long-term climate change plans are critical, 2008 through 2012 Protocol results show low reduction targets, attesting to their lack of long-term applicability and efficacy. And, as previously stated, the Protocol itself often lacked adequate incentivizing factors.

Replacement proposals (for long-term targets)

There are many post-Kyoto Protocol proposals for emissions reductions frameworks. Some involve bilateral agreements, while others call for regional cooperation in policy objectives. The Kyoto Protocol taught the world that global schemes that call upon a diverse set of actors could lead to overly broad, difficult to execute policies. To the extent that the policy substance is weak, reaching long-term greenhouse gas reduction targets becomes ever more difficult. And, while efficient environmental tax and emissions trading schemes can, depending upon the technology used, give rise to greater "know-how" and diminishing policy execution costs, long-term investment is still necessary, and hoped-for technological breakthroughs are never guaranteed.

One possible long-term policy is a joint-development protocol in which developed countries improve already existing devices or invent new technologies pertinent to climate change policy. The merit of this is that national interests align, and there is solidarity in motivation to participate and improve technology standards. Thus, over the long term, joint-development protocols are extremely useful for improving existing technology and inspiring new creations. A prime example is the cooperation between the U.S., Japan, Canada, the E.U., and Russia in their collective endeavor to establish the International Space Station. As seen here and elsewhere, joint-implementation projects perpetuate information sharing, among other benefits. And, while they do not guarantee widespread improvements in environmental cleanliness, they are still quite effective when used to supplement other frameworks.

APP is another exemplary process. It cuts both obligation and costs to correspondence, leading to greater, more reliable, easy-to-execute exchanges of information related to problems and operational challenges. This process, in turn, enhances new technology proliferation while also perpetuating research and development. And, it will surely be crucial to greater inter-sector correspondence and communication in the future.

At the same time, isolated, bilateral, and inter-regional agreements could potentially yield large differences among climate change policies. Furthermore, it is clear from the international Kyoto Protocol experience that a great amount of negotiation time is required when new frameworks are drawn. Even still, simply entering agreements without clear and meaningful targets discourages further aims to draft impactful policies.

SUMMARY

Since it did not set emissions reduction targets for developing countries, the Kyoto Protocol achieved limited success in alleviating the side effects of climate change. To this effect, a handful of "post-Kyoto" policies are on the drawing board. From here on, it is important to come up with incentives to participate and methods to ensure the efficacy of policies that aim to curtail global greenhouse gas emissions over the long run.

REVIEW PROBLEMS

1. Describe why international agreements are difficult.
2. Explain both the contributions and limitations of the Kyoto Protocol.
3. Outline the post-Kyoto Protocol options.

LEARNING POINT: EFFORTS MADE BY U.S. FIRMS

U.S. enterprises are showing greater support for emissions trading. Take for example, the United States Climate Action Partnership (USCAP), a cooperative organization of large U.S. enterprises and environmental protection organizations established in January 2007. It aims to promote legislation for greenhouse gas emissions reduction obligations hand in hand with the U.S. Federal Government policies.

USCAP membership is made up of over 30 American companies, including GE, Alcoa, BP America, and PG & E. And, with the participation of the 3 big automakers, USCAP has become even more capable of impacting the U.S. Federal Government.

USCAP's six proposal purpose topics are:

1. Investigating the causes of global climate change;
2. Recognizing of the importance of technological revolution;
3. Instituting efficiency that reflects the environment;
4. Creating advantages to economic opportunities;
5. Preserving justice that considers differing effects on each sector; and
6. Stressing prompt action.

Furthermore, the short to mid-term greenhouse gas reduction targets held by the USCAP include:

1. Limiting emissions increases for the next 5 years.
2. Reducing emissions by 10% of current levels over the next 10 years.
3. Reducing emissions by 20% of current levels over the next 15 years.

Moreover, the long-term target is to reduce emissions by 60–80% of current levels by the year 2050.

Just why is it that U.S. firms have taken it upon themselves to adopt these regulations? The reason is that applying different rules to different states could create an excessive burden for companies that are expanding their businesses across the country. Therefore, pledging to implement a unified system of rules across the country turns out to be ideal.

CHAPTER 5

Environmental value assessment

Chapter overview

This chapter outlines the methods for assessing the value of the environment using monetary units. In order to assess public works projects provided by the government, it is essential to compare the monetary effects and costs of environmental policies through cost benefit analyses of such policies. However, as there exists no market price for the environment, indicating the effects of environmental policies using monetary units is no easy feat. *Environmental assessment methods* were developed to achieve this end. While they are relatively new to environmental economics, environmental assessments grow increasingly important to the field overall.

The first task is to consider what, environmental assessments are. Out of the many existing means for determining environmental value, *willingness to pay* and *willingness to accept* amounts reflect monetary values based on economic decisions regarding the environment. Moreover, there are two main environmental assessment methods. The first involves observing economic behavior to produce indirect evaluations. The second involves making inquiries in direct assessments. This chapter details both methods and introduces environmental policy precedents that make use of them.

Chapter content

Section 5.1—This section draws from specific environmental problems as starting points for making environmental assessments. Furthermore, the role that willingness to pay (the maximum amount one does not mind paying) and willingness to accept (the minimum required monetary amount) have in determining environmental value are depicted.

Section 5.2—Here, methods assessing the environment using monetary unit measurements are detailed. The revealed preferences methods indirectly assess the environment by observing individuals' economic behavior. This section describes various revealed preferences method precedents and each of their special features.

Section 5.3—Stated preferences methods assess the value of the environment through direct inquiries to individuals. They draw attention to the value of the environment being left as it is, unused. This section introduces policy application precedents for this type of environmental assessment.

Section 5.4—Cost-benefit analyses of environmental policies compare the costs involved in implementing policy with the monetary benefits gained from the effects of the policy in order to indicate its efficiency. This section explains how this can be applied to public works.

Section 5.1: Environmental value

Utility and *non-use value*

Even in its simplest and most straightforward sense, "environmental value" is an extremely complex concept. By analyzing the ways it is utilized, the value of the environment can be broken up into two major categories: utility value and non-use value (see Figure 5.1.1). For example, consider the environmental value of a forest. Forests could be used as sources of lumber, or they could also be used for recreational activities like mountain climbing and camping. When a forest's wood resources are directly used in this way, it is said to possess *direct use value*. By contrast, when forests are enjoyed for their scenery, and they are indirectly used in a manner that does not lead to the loss of the forest or wood resources, they are said to have *indirect use value*. Moreover, when a forest is not used, yet can be used in the future, the value of leaving the forest for future use is known as *option value*. For example, tropical rainforests are home to plenty of plants and animals, and many of these species could be used to develop medical supplies for the future. However, in the event of deforestation, the potential to use such species for medical supplies is also lost. Therefore, rainforests could be considered to have *option value* in the sense that preserving them now (e.g. protecting the current *non-use value*) would generate future medical supply resources.

On the other hand, a forest could have value even if its resources go unused altogether. For example, the forests of Yakushima, Shirakami Sanchi, and Shiretoko are designated World Heritage Sites and exclusive habitats for rare species. Though their ecosystem resources remain untouched, their preservation is still important for future generations. The value of leaving the environment for future generations is known as *inheritance value*. Correspondingly, many believe that the threat of extinction faced by many species that inhabit preserved forests justifies maintaining *inheritance value* (even if no one were to gain from using ecosystem services like food, materials, etc.). The value that is yielded from the mere existence of wildlife in such scenarios is known as *existence value*.

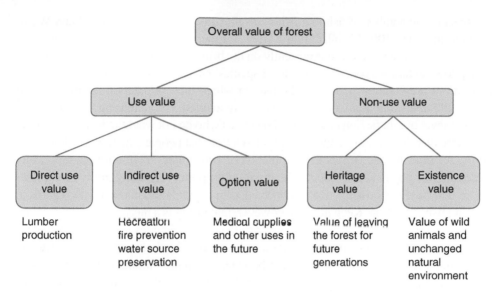

FIGURE 5.1.1 Forest value

While forests are valuable from the number of unique perspectives depicted, market prices only reflect the *direct use value* of lumber resources. Since market prices are not assigned to the other values, the overall value of forests goes underestimated.

Willingness to pay (WTP)

Now, the next question is, what can be done about the fact that some environmental values have no market prices and thus, the environment is improperly evaluated? In environmental economics, the *willingness to pay (WTP)* and *willingness to accept (WTA)* amounts are the scales used to measure the monetary value of the environment. WTP is the maximum amount of monetary contribution that one wouldn't mind paying regarding a change to the environment, and WTA is the minimum amount of monetary compensation that one would be open to receive for a change to the environment.

Table 5.1.1 Willingness to pay and willingness to accept amounts

	Environmental improvement	*Environmental degradation*
Willingness to pay	The maximum amount one does not mind paying to improve environmental quality	The maximum amount one does not mind paying to prevent environmental degradation
Willingness to accept	The smallest amount of monetary compensation one would demand upon conceding to the repeal of environmental improvement countermeasures	The smallest amount of monetary compensation one would demand for environmental degradation

Consider the number of fish in a fishery to illustrate an example of WTP and WTA. Assume there are 100 fish inhabiting a river. Furthermore, assume the fish species is native to this particular river, it is currently on the brink of extinction, and local residents living downstream are concerned about the species' survival. Next, assume the residents look into a preservation policy for the fish, in which their population numbers would rise to 200 and extinction would be successfully averted. By increasing the number of fish, the level of satisfaction among local residents who do not want the fish to go extinct also increases. In economics, the level of satisfaction that people experience is known as *utility*. Assume that the current fish population of 100 equates to 10 utility points, while increasing the population to 200 results in 20 utility points. By using utility as a guideline in this way, it is possible to observe changes in people's satisfaction with regards to averting fish extinction. Keep in mind, however, that utility is a measurement of satisfaction, not a monetary unit. Thus, it is necessary to convert changes in utility to monetary units, and in order to do so, one must examine the relationship between fish and currency.

Figure 5.1.2 indicates the relationship between a fish population and currency. The horizontal axis indicates the fish population levels, while the vertical axis indicates the income levels of residents. To the extent that fish population increases, or to the extent that incomes are high, utility increases. Therefore, the further up and to the right one proceeds on the graph, the higher the utility becomes. The curve on the graph is known as the *indifference curve*, and any points above this curve accrue an equal amount of utility. Assume that current conditions correspond to point *A*, where residents' income amounts to 5 million yen and the fish population is 100. However, when income drops to 4.99 million yen, the fish population reaches 200, represented by

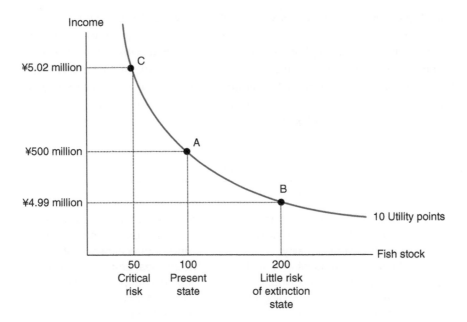

FIGURE 5.1.2 Indifference curve

point B on the indifference curve. While the utility does indeed increase in conjunction with a greater fish population, diminishing income cancels this out with an equivalent detraction from utility, and the utility ends up being equal to the value at point A. In the same manner, if income levels increased to 5.02 million yen, and the fish population decreased to 50, such conditions would be analogous to point C on the indifference curve. Moreover, while the decrease in fish numbers would lead to a reduction in utility, income increases could counteract this, resulting in unchanging utility levels.

Next, consider a fish preservation policy that aims to increase the fish population from 100 to 200 in order to prevent extinction. Figure 5.1.3 depicts the WTP and WTA levels for this sort of environmental restoration policy. There is a shift from point A to D, where income levels remain at 5 million yen but fish numbers increase from 100 to 200, resulting in an increase in utility from 10 to 20 points. Here, a decrease in income from 5 million yen to 4.99 million yen would reflect a shift to point B, but since points B and A are located on the same indifference curve, the utility remains unchanged at 10 points. In this scenario, the WTP is equivalent to the maximum amount of money one is willing to pay to increase the fish population size, depicted by BD, which is equivalent to 10,000 yen (or difference of 5 million yen – 4.99 million yen). If a monetary amount less than BD is paid (5,000 yen, for example), it would result in being above the 10-point utility line, and further payments would be necessary. Conversely, if a monetary amount greater than BD (20,000 yen, for example) is paid, then the utility would be less than the original utility at point A, meaning that too much is being paid. Finally, if an amount exactly equal to BD on the graph (10,000 yen) were paid, then the it is possible to remain on the indifference curve while restoring fish population levels and maintaining a utility amount equivalent to A. In this way, the WTP for environmental improvement

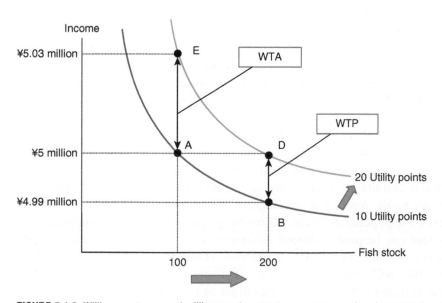

FIGURE 5.1.3 Willingness to pay and willingness to accept amounts (case of environmental restoration)

is equal to the maximum amount of money one doesn't mind paying to improve the environment.

Willingness to accept (WTA)

Next, consider the WTA compensation for environmental restoration. The previous example centered on policies aiming to preserve and subsequently increase fish populations from 100 to 200. At this point, assume that said preservation policy gets cancelled for financial reasons. This would result in a return to point A, representing a downturn compared to the transition to point D that had been realized previously. In turn, local residents object to this sudden cancellation of preservation policy, as utility drops from the 20 points present in the post-policy implementation scenario to the 10 of the current scenario. As an alternative to cancelling the policy, a subsidy equivalent to AD (or 30,000, derived from 5,030,000 yen − 5,000,000 yen) is paid out to residents. By doing this, there is a shift from point A to E, which is on the same indifference curve as point D that had been reached after implementing policy, rendering utility equivalent to the post-improvement 20-point levels. This effectively means that residents would allow the previous policy to be abolished if they were paid compensation equivalent to AE, their WTA amount for the fish population decreases. If the WTA compensation amount were less than AE, then the utility would be less than point E and less than the 20 points realized after preservation policy is implemented, so locals would not consent. Conversely, if the WTA compensation amount were greater than AE, then the utility would also be greater than point E, and the utility would be greater than the 20 points of point E, which would mean that such a high level of compensation would be unnecessary. In this way, the WTA compensation is equivalent to the smallest amount of money that is necessary to compensate for cancelling environmental restoration policies.

Previous examples depict WTP and WTA compensation for environmental restoration. From here, consider the opposite case of environmental degradation.

Assume in Figure 5.1.4 that due to development, the forests surrounding river are cut down, and what results is the fish population drops from 100 to 50, where it is on the verge of extinction. In this case, also assume that citizen utility drops from 10 points to 5 points. At this point, it is only necessary to compensate citizens with monetary amounts equivalent to CF on the graph (or 20,000 yen, derived from 5,020,000 yen − 5,000,000 yen) in order to preserve the same conditions as before the fish population diminished. In this way, the smallest amount of compensation for environmental degradation is equivalent to the WTA compensation amount for environmental degradation. Conversely, one could imagine the residents using their money to buy up the forests surrounding the river in order to disrupt development and preserve fish population levels. AG on the graph depicts the maximum amount of money that can be paid to disrupt development. If the citizens attempt to pay more than this amount, then the fish population would decrease alongside development and utility will decrease to 5 points, meaning that such payments would be impossible. It follows, then, that the maximum amount that residents pay to deter environmental degradation is their WTP.

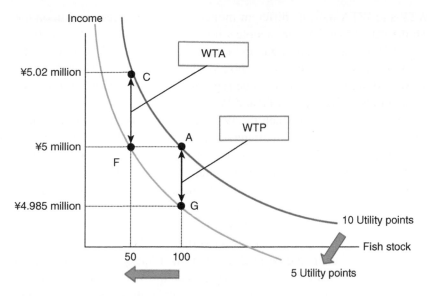

FIGURE 5.1.4 Willingness to pay and willingness to accept amounts (case of environmental degradation)

Special features of WTP and WTA

Environmental economics uses WTP and WTA as the means to evaluate environmental value, but why do the WTP and WTA correspond to environmental value? The following summaries depict the special features of WTP and WTA amounts:

1. They evaluate environmental degradation.
2. They evaluate changes in utility.
3. They are different depending upon the individual.
4. They can become various monetary amounts.

First, WTP and WTA amounts are defined by changes to the environment, whether they are improvements or degradations, and it is important to note that current environmental conditions alone cannot be used to determine the monetary value of the environment. For example, if the current fish population is 100, one cannot determine the value of these 100 fish. When there are no changes to the environment that involve two or more comparative conditions (e.g. fish numbers increasing from 100 to 200 or decreasing from 100 to 50), WTP and WTA cannot be defined.

Secondly, WTP and WTA are monetary amounts that correctly reflect changes in utility that arise through changes to the environment. Willingness to pay and WTA increase to the extent that utility rises. When fish population levels increased from 100 to 200, 300, and 400 in Figure 5.1.3, WTP and WTA changes allude to the fact that utility rises to the extent that fish populations increase, and both WTA and WTP also increase accordingly. Thus, one could view WTP and WTA as conversions of utility changes to monetary units.

Thirdly, WTP and WTA indicate different monetary amounts depending upon the individual. To the extent that there are people who believe in protecting the environment, WTP and WTA amounts are high, and conversely, once could imagine WTP and WTA levels equivalent to 0 for those who have absolutely no concern about the environment. Accordingly, the sense of value regarding the environment at the individual level is said to be reflected in WTP and WTA amounts.

Finally, there are many cases in which the WTP and WTA are different monetary values. While it is true that WTA and willingness reflect individual value perceptions of the environment as stated above, it is not necessarily true that they will both be the same monetary amount. For example, in a scenario where wildlife could become extinct due to development, there is nothing that can be done to reverse extinction, so there would probably be quite a few people who believe that no amount of money that could be received is worth permitting extinction. In this case, the WTA compensation amount is infinite, but the WTP amount will never surpass one's own income, regardless of how much they think should be paid, so there is a large divergence yielded between the WTA and WTP amounts. In this way, since the WTA amount becomes an excessively large amount of money, when environmental policy is actually utilized in reality, WTP is generally adopted.

In the previously mentioned ways, the WTP and WTA amounts possess effective special features for evaluating the monetary unit value of the environment on a linear scale, but they cannot be used to acquire market data in the same way that market prices can. Because of this, it is necessary to either estimate indirectly people's WTP and WTA amounts based on economic activity, or directly ask and subsequently calculate peoples' WTP and WTA amounts.

SUMMARY

The value of the environment includes use and non-use values, but outside of direct use value, market prices do not exist for other environmental values. Thus, from this, the price of the environment fails to reflect the true value of the environment. In environmental economics, WTP and WTA are used when determining environmental value. WTP and WTA properly reflect changes in utility that arise with changes to the environment, and it is necessary to utilize special evaluation methods for estimating the WTP and WTA amounts of these linear monetary measurement that reflect environmental value perceptions of individuals.

REVIEW PROBLEMS

1. Assume the following utility function $U = (q \times M)/100$. Importantly, q represents the fish population (numbers), and M represents income (tens of thousands of yen, $M = 100$).

    a. State the WTP amount when fish populations increase from 10 to 20.

    b. Assume the fish population decreases from 10 to 2 due to river pollution. State how much the WTP is in order to preserve the current fish population level at 10.

    c. There is a preservation policy in place to increase the fish population levels from 10 to 20, but it is cancelled. State how much the WTA compensation is in such a scenario.

2. Indicate an evaluation target when the WTP and WTA are comparatively close, as well an evaluation target when there is thought to be a large gap in the WTP and WTA.

LEARNING POINT: DIVERGENCE BETWEEN WTP AND WTA AMOUNTS

Even for the same assessment object, large discrepancies in assessment values come about depending upon whether the WTP or WTA amount is being queried. Generally speaking, the WTA value tends to be greater than the WTP amount. One origin of this is what is known as the substitution effect. To illustrate this, consider examples where tap water quality declines and wildlife extinctions occur.

In the case of tap water, when consumers receive money to purchase purifiers or bottled mineral water, they can avoid water quality concerns. Since a replacement exists, then, the indifference curve becomes a nearly straight line (as depicted in Figure 5.1.5), and the WTP and WTA amounts are quite similar. On the other hand, in scenarios involving wildlife, no amount of money can artificially replace or resurrect living creatures once they have gone extinct. This means that because these private goods cannot be replaced, regardless of how much is paid, nothing exists which covers the cost of wildlife extinction. What results is a concave indifference curve as shown in the diagram. The exceptionally large WTA amount greatly diverges from the WTP. In this manner, the existence of a substitute affects the diverging or converging relationship between the WTP and WTA amounts.

Section 5.2: Environmental valuation method 1: revealed preferences method

What are environmental evaluation methods?

Since there exists no market price for the environment, it is essential to come up with special evaluation methods for measuring the monetary unit price of the environment. These measures are known as *environmental valuation methods*. While a

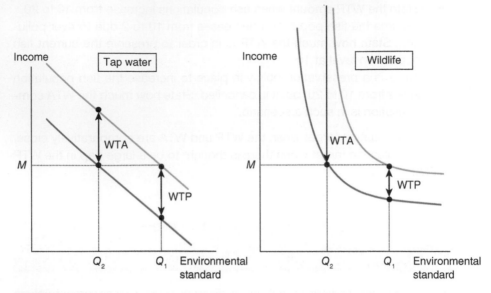

FIGURE 5.1.5 Divergence between willingness to pay and willingness to accept amounts

number of environmental valuation methods have been developed, they are largely classified into two groups: revealed preferences and stated preferences methods. *Revealed preferences methods* are techniques for indirectly placing value on the environment by observing how much the environment affects people's economic behavior. Conversely, stated preferences methods (refer to Chapter 5, Section 5.3) are measures for evaluating environmental worth through direct inquiries to individuals.

Table 5.2.1 indicates the special features of each revealed preferences method. The methods include the *substitution*, *travel cost*, and *hedonic* methods. The *substitution method* makes evaluations based on the cost of replacing an aspect of the environment with a private good. For example, when evaluating the water source conservation function of the environment, one could examine how many man-made dams would be required to perform a given forest's natural water source conservation function and ascribe to the environment monetary value equal to the dam construction cost. The substitution method is direct and easy to understand, and since it also allows for values to be compared relatively simply, it is often used in preliminary research of environmental evaluation methods. For example, in 1972, the Forestry Agency ascribed 13,000,000,000,000 yen in value to diverse functions of forests across the country through the substitution method. Furthermore, since 1982, the Ministry of Agriculture, Forestry, and Fisheries has been making evaluations through the substitution method, and in 1998, it indicated the total value of all farmland across the country to be worth 7,000,000,000,000 yen.

While the substitution method has been frequently used in such a manner, its drawback is that it cannot make evaluations when there are no suitable private goods with which to derive value and appropriate to the target of evaluation (e.g. the environment.) For example, when evaluating wildlife that is on the brink of extinction, it is

Table 5.2.1 Special features of each revealed preference method

Classification	Revealed preference method		
	Indirectly ascribe environmental value by observing human behavior		
Title	Substitution method	Travel cost method	Hedonic method
Contents	Evaluate based on the cost of replacing a good with a separate private good	Evaluate based on travel cost to a destination	Evaluate based on the environment's impact on rents
Applicable scope	Usage value Limited to such pursuits as water quality restoration and landslide prevention	Usage value Limited to recreation, sight-seeing, and other travel-related matters	Usage value Regional amenities, water pollution, noise pollution, death risk
Advantages	Intuitive and easy to understand	Not much information is necessary Only requires travel cost and visitation frequency information	Information gathering cost is low; rent and wages can be acquired from market data
Drawbacks	Evaluations cannot be made if private good exists to serve as a reference	Limited to the scope of recreation	Cannot evaluate what exists outside of ideal markets; must assume ideal markets are perfect markets
Precedents	Evaluations of forests, farmland, and other areas with diverse natural functions, effects of water source development	National park improvements, city park improvements, green space improvements	Air pollution policy, construction damage policy, home improvement

difficult to find appropriate, substitutable private goods. Perhaps the cost of feed or fodder that would be paid by a zoo for a given creature could be used in evaluations, but it is difficult to ascribe the same value to animals eating in zoos and animals that inhabit natural settings. Similarly as there exist no private goods that are of the same worth as wildlife diversity or the ecosystem, the substitution method cannot make evaluations in such cases. As the social concern for global environmental issues such as tropical rainforest destruction and climate change have heightened since the latter half of the 1980s, modern interest in the substitution method has greatly lessened due to its inability to evaluate such global environmental issues.

The *travel cost method* is a technique for evaluating the price of recreation based on how much it costs to travel. While the travel cost method cannot be used to evaluate the prices of such factors as ecological diversity or the ecosystem, it is often used in recreation related policy evaluations, such as national park maintenance.

The *hedonic method* is a technique for evaluating the environment based on the influence that the environment has on land prices and rent costs. The hedonic method can evaluate market data such as land prices and rents, and because it is easy to acquire the necessary information for making evaluations, it is used in policy assessments of noise and air pollution. However, it is limited to evaluations within land and labor markets. For example, because climate change yields impacts on the global scale, regardless of where one lives domestically or what profession they pursue, the impact they receive should be equal. Accordingly, the result of climate change policy does not reflect land prices and rents, making it difficult to be used in such a scenario.

Below, the travel cost and hedonic methods, two model revealed preferences methods, will be explained in detail.

Travel cost method

The travel cost is a method for evaluating the price of recreational activities based on the relationship between the cost of travel to a destination and number of visits (otherwise known as the visit rate). For example, many tourists from across Japan visit the Shiretoko World Heritage Site in Hokkaido x number of times because of the value that they personally perceive it to have. One fundamental tenet of the travel cost method is that the travel cost that tourists pay must be reflected in the visitation prices of destinations like Shiretoko.

Figure 5.2.1 depicts the travel cost method. Consider the travel price to a national forest. The vertical axis indicates the price of traveling to the public forest, while the horizontal axis depicts the visitation frequency. The relationship between the travel cost and visitation frequency are depicted by the demand curve D on the graph. When the travel cost is high, the visitation frequency is low, so the demand curve has a down-rightward trend. The demand curve can be estimated from the travel cost and visitation frequency of this national park's tourists.

Assume that the round-trip cost from home to the park for a given visitor is 1,000 yen. This particular visitor makes a park trip 30 times in a year. According to the graph, the visitor would not mind paying up to 1,400 yen on the tenth visit, but in reality, they are only paying 1,000 yen, so the remaining 400 yen remains in the visitor's pocket. Since the visitor would pay 1,200 yen on the twentieth visit, the visitor would gain an extra 200 yen. Upon the thirtieth visit, the amount of travel cost that the visitor would pay and the actual travel cost are both 1,000 yen, so there are no gains earned by the traveler.

Regarding total visitation numbers, the shaded area on the graph becomes the earnings of the travelers. This shaded area is known as the consumer surplus. In the example depicted in the graph, the calculated consumer surplus is the shaded triangle area that amounts to 9,000 yen (low area 30 × height 600 / 2). Because the consumer

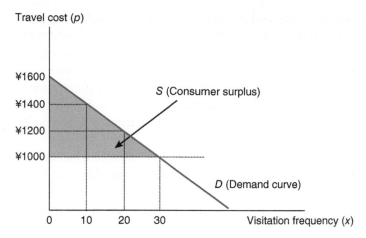

FIGURE 5.2.1 Travel cost method

surplus is the amount pocketed by the traveler when visiting the national park, it can be viewed as a representation of the visitation value of the national forest.

By estimating the demand curve using the visitor's travel cost and visitation number information in this way, the visitation value can be calculated. Next, assume a promenade is added to the national forest, and visitors can look forward to enjoying the forest scenery. The appeal of the national forest rises with the addition of the promenade, and as depicted in Figure 5.2.2, the national forest's demand curve shifts to the right. The people who had been visiting 30 times per year up until now will increase their visitations to 35 times because of the addition of the promenade. After the creation of the promenade, the consumer surplus increases by the lightly shaded area depicted on the graph, meaning the addition of the promenade led to an increase in consumer surplus.

In this way by making use of the relationship between the travel cost and visitation frequency, the travel cost method can measure both the visitation value of such places as metropolitan and natural parks as well as the concomitant effects of park

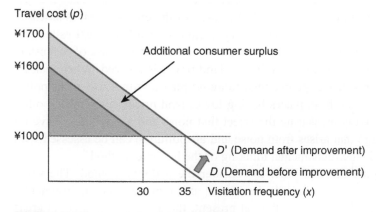

FIGURE 5.2.2 The impact of park improvements

maintenance of travel destinations. Beyond this, it is indeed true that unlike the given example which provided the yearly visitation rates to a nearby park, it would make sense that the yearly visiting frequency and visitor numbers of such far away national parks as Shiretoko are quite few. Investigations into where the visitors came from, including Hokkaido, Tohoku, Kanto, and other regions, all lead to different travel costs and visitation frequencies. Those who come from afar have high travel costs and low visitation frequencies, so the down-rightward trending demand curve can be estimated by making use of the visitation rate.

The advantage of the travel cost method is that it can make assessments by using only travel cost and visitation frequency data, but on the other hand, it has a few drawbacks. The first is that the evaluation scope of the travel cost method is limited to aspects related to recreation. The travel cost cannot assess the worth inherent in specifics that are not visited, such as the non-use value of biological diversity and the ecosystem.

The second is that calculating travel and opportunity costs is inherently arbitrary. For example, there is the chance to earn income if the travel and visitation time period allowed to travel were used for labor. Since these time-based opportunity costs are yielded, when travel costs are calculated, not only the actual expenditures of train and gasoline prices, but also the opportunity costs based on time must also be calculated. To calculate opportunity costs, there are many methods, from using the income rate as it is, to using half of the income rate, to using a zero opportunity cost, meaning that in reality, there is no one-size-fits-all approach.

The third drawback is that it is essential to consider the effects of substitutability. For example, when a forest park is shut down due to local infrastructure or housing development, a traveler might choose to go to another forest park in the area, but if this kind of substitution behavior is ignored, the effect of the park closing is over assessed. In order to consider these substitutable areas, a multi-site model that considers multiple destinations is being developed, and it is commonly used in many modern travel cost method analyses.

Hedonic method

The hedonic method is a process by which the value of the environment is assessed based on the effect the environment has on representative markets. Real estate and labor markets are often used as such representative markets. In the case of real estate markets, the effect that the environment has on land prices is utilized. When making decisions about their residence, people tend to avoid places with noise or air pollution problems, resulting in those places having lower land prices compared to other areas. Thus, by effectively measuring the effect that noise and air pollution have on land prices, the damage that arises from noise and air pollution can be assessed with monetary units. This is the fundamental thinking of the hedonic method.

Take a look at Figure 5.2.3 in order to consider the hedonic method. The vertical axis represents land prices, while the horizontal axis represents an environmental standard such as air quality. Assume that, at present, there are two residential areas *A* and *B*. Residential area *A* is one in which air pollution is quite severe, while on the other hand, residential area B has clean air. In such a circumstance, people would avoid area A and its heavy air pollution, rendering the land price for area A, P_A, less than the land price for area B, P_B. In this way, to the extent that air quality is improved, land prices increase, so the relationship between air quality and land prices becomes an up-rightward trending curve as depicted in the graph. This is known as the hedonic price curve. Here, let's assume that air pollution policy is implemented in order to improve the air quality of area A to the same standards as level B. When the air quality improves from Q_A to Q_B levels, the land price amounts rise by $P_B - P_A$. With the hedonic method, this amount by which land prices increase is assessed as the result of air pollution policy.

Next, take a look at the relationship between the assessed price of the hedonic method and the WTP amount (Figure 5.2.4). To simplify, let us assume only one supplier of residential space. The indifference curve of Resident 1 is U_1 on the graph, while that of Resident 2 is U_2. Since both the highest possible environmental quality and cheapest possible land prices are both most desirable, the further we proceed down and to the right on the graph, the greater the resident's satisfaction level (utility) becomes, and the indifference curve of such reflects this in its shape on the graph. At this time, Resident 1 prioritizes cheap land prices over environmental quality, leading them to choose residential area A. At point A, resident 1's indifference curve U_1 connects with the hedonic price curve, and other points above the hedonic price curve are all to the up-right side of the U_1 indifference curve, so resident 1's utility is maximized at point A. Conversely, Resident 2 prioritizes air quality over land price and chooses point B.

Here, assume the environmental quality of residential area A is improved from Q_A to Q_B. As previously stated, by assessing with the hedonic method, the amount by which land prices increase as a result of environmental policy can be expressed by $P_B - P_A$. However, as the point C, not point B, is on the same indifference curve as point A, the maximum amount that resident one can pay for this environmental policy (WTP) is equivalent to $P_C - P_A$. In other words, the valuation amount that results from the hedonic method is maximized not at the WTP amount, but only at $P_C - P_A$. If all of the

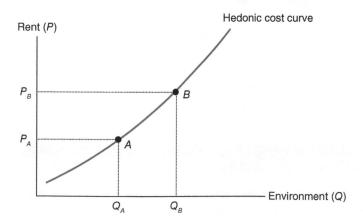

FIGURE 5.2.3 Hedonic cost curve

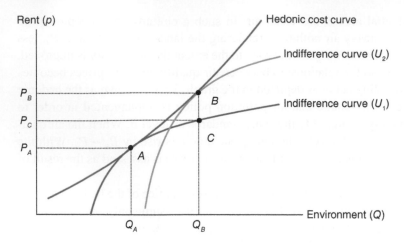

FIGURE 5.2.4 Hedonic method valuation and willingness to pay amounts

residents had the same preferences, and their indifference curves were unified, then the amount assessed by the hedonic method and the WTP amount would be uniform. Yet, if each citizen had different preferences as in the graph, the hedonic method produces the maximized evaluation. There, after estimating the hedonic price curve, there is also a secondary estimation for special estimations via shifts in the indifference curve due to differences in individuals, but because individual data of residents is required, in reality, in most cases the hedonic curve is used as it is to make assessments.

When the hedonic method only uses land prices and attributes for residential markets, incomes and professions for labor market assessments, its advantage is the ease of acquiring necessary information. However, the following two drawbacks exist within the hedonic method. First, it requires the assumption of a perfectly competitive market, yet there are many regulations and practices among land and labor markets, and furthermore, when choosing residential areas, moving and other transaction costs are borne, meaning it is difficult to imagine that the perfect competition assumption holds true. Secondly, the hedonic method cannot assess environmental values that exist outside of particular ideal markets. For example, in the case of climate change, regardless of where one lives in the world, or what their profession is, the effects of climate change will be the same, so the hedonic method that utilizes land and labor markets, for the effects of climate change cannot be assessed. Hedonic methods centered on land markets can only assess the environmental work of living on land in a specific region, but they cannot assess the global environmental value of climate change problems.

SUMMARY

Revealed preferences methods are measures for assessing the value on the environment by estimating the effects that the environment has on people's economic behavior. The travel cost method assesses the recreation value based on the relationship between the travel cost and visitation frequency (or the visitation rate). The hedonic method evaluates death risk, air pollution, and noise pollution by estimating the effect the environment has on land prices and incomes. Although the assessments among revealed preferences methods require information that is simple to acquire, their drawback is that they are unable to assess non-use values.

REVIEW PROBLEMS

1. The following recreation demand function is for a beach: $x = 4 - (p/500) + q$. The visitation amount is represented by x (times), the travel cost is represented by p (yen), and the water quality is represented by q (points).

   a. Assume current water quality is 0 ($q = 0$). Express a given traveler's consumer surplus if their travel cost were 1,000 yen.
   b. Assume that when water quality improvement policy is executed, water quality increases by two points, resulting in $q = 2$. Express how much the water quality improvement policy was for the traveler.

2. The utility function of residents exists when the relationship between air quality (q points) and land prices (p yen) is represented by $p = 5q$. Assume it is $u = q(20 - p)$.

   a. Express the extent to which these residents would prefer air quality.
   b. Assume that air quality is doubled compared to current levels. Express the extent of this effect using the hedonic method assessment.
   c. Express the degree to which the assessment is maximized compared to the WTP amount.

LEARNING POINT: THE U.S. CLEAN AIR ACT

In the U.S., economic evaluations are of air quality are performed through the *Clean Air Act.* The U.S. EPA performed comparisons of the benefits and costs borne since the implementation of air pollution countermeasures in the *Clean Air Act*. Then, the health damage prevention effect of the act is assessed with the value of statistical life (VSL, refer to Chapter 6, Section 6.3 for details). The VSL is defined as the WTP for decreased risk of death divided by the scope of the risk reduction. Past EPA VSL assessment precedents of 26 studies (21 of which were studies which used the hedonic method, while 5 of them used CVM) applied an average value of $4.8 million to perform benefit assessments for avoiding single deaths. The result was that in 1990 values, the $500 billion *Clean Air Act* cost amounted to $22.2 trillion in benefits (90% confidence interval $5.6 trillion – $49.4 trillion).

Section 5.3: Environmental valuation method 2: stated preferences method

What are stated preferences methods?

Stated preferences methods are those that place monetary unit values on the environment by directly asking people what it is worth to them. Since the revealed preferences methods (see Chapter 5, Section 5.2) ascribe value to the environment indirectly by assessing the impact that the environment has on economic behavior, they are incapable of evaluating the non-use value of wildlife, the ecosystem, and other factors that do not have an impact on human behavior. Conversely, stated preferences methods make it possible to assess non-use value that is not reflected in human behavior by asking people directly. In the early 1990s, general awareness of climate change, the loss of biodiversity among wildlife, and other global environmental problems had risen across the board, yet many of the global environmental issues were rooted in non-use value and could not be assessed without stated preferences methods. Therefore, the early 1990s witnessed an increased focus on stated preferences methods.

Stated preferences methods include the *contingent valuation method (CVM)* and *conjoint analysis* (see Table 5.3.1). Through CVM, the value of the environment is assessed by stating an ideal environmental policy method and directly asking individuals about their WTP and/or WTA levels regarding environmental changes. CVM has a broad scope for assessments, from use values derived from recreation and sightseeing, to non-use values of wildlife and the ecosystem. However, since it employs questionnaires, there is the risk of question content giving rise to bias among the answers. And, if the questionnaires are not carefully crafted, the assessment values could prove to be unworthy in the end. CVM has been used in many environmental policies around the world. It has proven particularly useful as a means to calculate damage costs of environmental degradation by judicial courts seeking to determine reparation amounts after an infamous tanker accident in the U.S. Events of this sort drew the attention of the international community to CVM (see this section's **Learning point**).

Conjoint analysis highlights noteworthy proposals from several environmental policy candidates and evaluates the environment through policy preferences. Unlike CVM, its special feature is that it can separate the content of environmental valuations. For example, the value of the environment includes wood production, recreation, water source preservation, wildlife preservation, and other factors. With CVM, it would be difficult to separately assess the value of each individual aspect, but through conjoint analysis, aspects of environmental value can be assessed individually and distinctly. Since both conjoint analysis and CVM employ questionnaires, they both run the risk of taking on bias. Conjoint analysis is the newest evaluation method, and due to a lack of legitimizing research, it has been practically applied to but a few environmental policies.

Table 5.3.1 Details of stated preferences environmental assessment methods

Classification	Stated preferences method	
	Assess environmental value through direct inquiries regarding environmental worth	
Title	CVM	Conjoint analysis
Details	Evaluate through inquiries of one's willingness to pay or willingness to accept amounts for changes to the environment	Evaluate through inquiries of one's preferences after indicating numerous possible environmental policies
Applicable scope	Use as well as non-use values Extremely large scope, including recreation, sight-seeing, wildlife, diversity of species, and the ecosystem	Use as well as non-use values Extremely large scope, including recreation, sight-seeing, wildlife, diversity of species, and the ecosystem
Advantages	Wide range of applicability Existence value, heritage, and other non-use values can also be assessed	Wide range of applicability Environmental worth can be analyzed and assessed using unit measurements
Drawbacks	Large costs involved in collecting information though questionnaire surveys Bias is common	Large costs involved in collecting information though questionnaire surveys Bias is common
Practical examples	Recreational facility improvements, wildlife protection, ecosystem preservation, climate change countermeasures, tropical rainforest conservation	There are few precedents of this being applied to environmental policies

CVM

Through CVM, ideal environmental policies are stated, and the WTP and/or WTA amounts for environmental improvement or degradation are directly inquired from individuals in order to assess environmental value. For example, consider a case where CVM is used to assess the value of a forest (Figure 5.3.1). Currently, 10 hectares (ha) of the forest are preserved as wildlife habitat. However, assume that as development of the surrounding regions poses environmental risk, many begin to consider policies in which the preserved area is increased from the present 10 to 20 ha. To assess this preservation policy using CVM, first, CVM questionnaire respondents must be introduced to the current situation that includes a 10 ha preservation area. Then, respondents must be informed that the policy aims to increase the preservation area to 20 ha. Finally, individual respondents must be asked how much they would be willing to pay to realistically carry out such a policy. The WTP amounts are asked from ordinary households, so if the WTP amount is multiplied by the number of households to be affected by the policy, then the aggregate value can be attained.

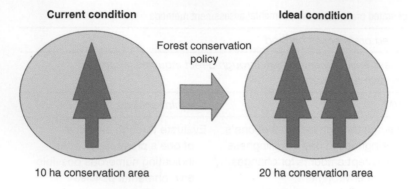

Current condition   **Ideal condition**

Forest conservation policy

10 ha conservation area   20 ha conservation area

How much would you pay for this forest conservation policy?

FIGURE 5.3.1 CVM scenario (environmental restoration)

Furthermore, WTP amounts can be assessed even in scenarios where the environmental quality is decreasing (Figure 5.3.2). Assume that the current 10 ha forest preservation area is to be revised, and forests will in the near future be cut down for housing development. In this case, respondents would be informed of the housing development plan in question, and then they would be asked how much they would pay to prevent its execution.

Similarly, CVM assesses environmental value through the following three processes:

1. Expressing the current environmental condition.
2. Expressing or estimating the environmental condition after some sort of change.
3. Asking for the WTP to remedy or nullify the environmental change.

It is essential to indicate not only current but also future environmental conditions after changes have occurred. As explained in Chapter 5, Section 5.1, WTP is tied

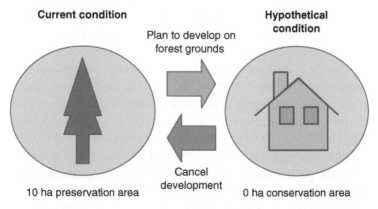

Current condition   **Hypothetical condition**

Plan to develop on forest grounds

Cancel development

10 ha preservation area   0 ha conservation area

How much would you pay for this forest conservation policy?

FIGURE 5.3.2 CVM scenario (preventing environmental degradation)

to improvements or downgrades, each of which represent particular environmental changes. Thus, WTP cannot be determined using assessments of present conditions alone. It follows, then, that simply defining the current 10 ha preservation area and inquiring how much individuals would pay would not allow one to obtain an accurate WTP value.

The questions within CVM surveys stem from a central idea: determining a WTP amount. To this end, a variety of question styles have been developed and polished over time (see Figure 5.3.3). Early CVM examples allowed respondents to freely fill in monetary amounts in what is known as the *free response format*, while other auction style precedents portrayed ascending price options in the *attached value game format*. Unfortunately, a number of invalid responses often appeared (e.g. radically high, radically low, or no response) in the former, while the latter encouraged bias with the monetary amounts it purported, rendering them both ineffective and largely unused in modern times.

The *pay card format* is one in which a card with price choices already filled in is used, and the individual can choose which WTP amount closely matched their personal valuations. Since it only involves choosing from a list of choices, there are few invalid responses, and there are also question format sheets that can be mailed in, but there is potential for the scope of the price selections to influence the actual responses. The *two-choice selection format* indicates a monetary value and leads respondents to choose "yes" or "no" based upon whether or not it is agreeable to them. Choosing between "yes" and "no" places a relatively low burden on the respondent, and it is an easy question type to respond to. Since the two-choice answer format includes a question style that is related to everyday consumer behavior such as making or not making a purchase based on the price of a product, it brings about comparatively little bias and is used the most nowadays.

The two-choice selection format only allows for "yes" and "no" data related to the stated prices to be acquired, so more statistical analysis is needed to estimate the WTP amounts. Figure 5.3.4 indicates a WTP estimation method. The horizontal axis indicates the monetary amount, and the vertical axis represents the "yes" response rate. With the two-choice selection method, a diverse set of prices, such as 200, 500, 1,000, 2,000, 4,000, 8,000, and 10,000 yen are stated, and each respondent is randomly shown one of the prices. Then, for each indicated value that "Yes" rate among respondents is searched for and filled in on the graph. When the majority of respondents choose yes when the price is low, the yes rate goes lower when the price rises. In the case of this graph, over 90% of respondents said yes when the price was 500 yen, but only 40% did when the price was 4,000 yen, and the yes rate dropped to less than 10% when the price reached 10,000 yen.

However, using statistical analysis of the curve that fits each point, the decay curve like the one in the graph can be derived. Here, where the yes and no responses each make up half, we can see the median value of WTP. On the other hand, by calculating the area beneath the decay curve, the average value of WTP can be derived. Generally, when calculating the area under the curve, the part up until the largest indicated value is calculated.

(1) Free response

> How much would you be willing to pay to designate this forest as a conservation area and protect its ecosystem? Please fill in a monetary amount in the space provided below.
>
> ¥

(2) Price bid game format

> Would you pay   ¥500 to designate this forest a wildlife reserve to protect its ecosystem?
>   「Yes」
> Would you pay ¥1,000?   「Yes」
> . . .
> Would you pay ¥9,000?   「No, I would not pay that much.」
> In that case, would you pay ¥8,500?   「Yes, that seems reasonable.」

(3) Payment card format

> How much would you be willing to pay to designate this forest as a wildlife reserve to protect its ecosystem? Please circle an answer from the choices below.

1.	¥0	2.	¥100	3.	¥300	4.	¥500
5.	¥800	6.	¥1,000	7.	¥2,000	8.	¥3,000
9.	¥5,000	10.	¥8,000	11.	¥10,000	12.	over ¥20,000

(4) Binary choice format

> Would you pay ¥1,000 to designate this forest as a wildlife reserve to protect its ecosystem? Please select one answer from below.
>
> 1. Yes   2. No

FIGURE 5.3.3 CVM question formats

Bias

In this way, CVM uses questionnaires to indicate both the current and post-change environment, and doing this to ask for a WTP is how environmental value is assessed. However, in scenarios where questionnaires are inappropriately used, there is potential for bias to arise and undermine the reliability of the environmental price. It is known from experimental studies up until now that CVM could potentially yield various forms of bias. Below are the most common types of bias that can arise.

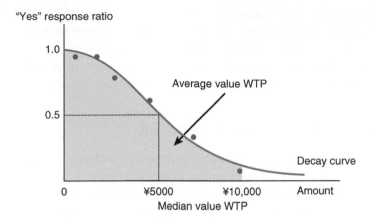

"Yes" response ratio

FIGURE 5.3.4 Method for estimating willingness to pay amount (binary choice format)

Strategic bias

The phenomenon in which respondents intentionally purport evaluations that are excessively small or excessively large is known as the strategic bias. For example, assume that an environmental preservation policy is already decided in advance and the responded price is not a cost to preserving the environment. In this scenario, if the responded price were actually levied from the respondents, then in order to evade the financial burden, respondents would purport a price that is lower than their actual WTP amount. Conversely, assuming the actual financial burden is not required from respondents, then dependent upon the responded price, environmental preservation policy could be implemented or discarded. In this scenario, since the respondent does not have to worry about upholding a financial burden, they will likely report a price higher than their actual WTP. In this way, strategic bias is a phenomenon in which respondents intentionally lie about prices in their responses. Moreover, the strategic bias tends to be prevalent in the free response formats where prices are freely given by respondents, yet in the two-choice answer formats where respondents are asked whether or not they will pay an indicated price, the strategic bias does not arise.

Bias based on information that will trail the evaluation

Among respondents, confusion based upon question content that is unfamiliar among CVM, and in order to search for where an answer does not hold the respondent accountable could be a source of bias. For example, in the price card format, it is possible for the scope of the indicated price to exert influence on the respondents, and this phenomenon is known as the *scope bias*. For example, in the case where one choice is elected from 0 to 3,000 yen, as well as in the case where a price between 0 to 10,000 yen is selected, the response results will probably vary. This is yielded from the respondents' recognition of the scope of the indicated price as a comparative reference point for evaluation. In order to get rid of this kind of bias, when question topics are being created, one should really keep in mind that to the greatest extent possible question content should be easy to respond to. And, at the same time, information that could potentially "trail" the respondent (especially information related to price amounts) must be excluded from question content.

Bias that arises from communication mistakes

CVM asks for WTP after conveying to the respondent current and post-change environmental conditions. However, if the information supplied to the respondent is not in line with the overall aims of the questioner, then "scenario communication mistakes" could potentially arise. There are many different types of scenario communication mistakes.

The first is when unrealistic scenarios or scenarios that are not theoretically appropriate are utilized. For example, in order to prevent global warming, in the case where a scenario for reducing greenhouse gas emissions by 50% of current levels is given, the responses are rejected as unfeasible, or in other cases, true responses will likely not be provided.

The second mistake arises when the scope of the evaluation target is not properly conveyed. For example, in the case of forest preservation, some respondents probably imagine tropical rainforests, while others likely imagine all forests across the globe. This kind of bias through mistaking the scope of an evaluation target is known as the "Whole/part bias."

The third bias arises from inappropriate payment processes during inquiries about WTP amounts. People often host feelings of rejection towards taxes that arise as financial burdens, but conversely, people feel satisfied by contributing money to some sort of foundation. This effect that payment processes has is known as the "payment process bias."

The scenario communication mistakes like those mentioned above result from the questioner not properly conveying their aims and intentions to the respondents. In order to resolve these issues, it is imperative to make use of photos and images that appropriately convey the intentions of the surveyor to respondents while also making sure that the aims of the surveyor are clearly expressed in preliminary questionnaires.

SUMMARY

Stated preferences methods are those by which the value of the environment is assessed by directly inquiring to individuals about environmental value. CVM indicates present as well as supposed environmental conditions, and makes assessments by inquiring the WTP for particular environmental changes. Conjoint analysis makes assessments by asking for agreeable replacement policies for present environmental policies. CVM and conjoint analysis are two of the few methods that can evaluate non-use values, but since they utilize questionnaires, they could potentially yield bias. Environmental assessment methods are not only performed with technological research, but must be utilized from multiple perspectives, including in actual environmental policy.

REVIEW PROBLEMS

1. Discuss problems with a CVM survey that uses the following to inquire about WTP:

   "A forest is to be preserved. How much do you mind paying in order to preserve this forest?"

2. Using the previously mentioned survey and the binary choice selection method, collect the following data:

   a. Create a scatter plot graph on a grid with the yes response rate on the vertical axis and the price on the horizontal axis. Then, connect each point with a straight line.
   b. Upon approximating the decay curve using this straight line, determine the median WTP.
   c. Upon approximating the decay curve using this straight line, determine the average WTP. For the area calculation, include the maximum revealed amount.

LEARNING POINT: DAMAGE ASSESSMENT OF THE *EXXON VALDEZ* TANKER OIL SPILL

On March 24th, 1989, the *Exxon Valdez* tanker traveling through the Prince William Strait in Alaskan waters ran aground, leading it to spill upwards of 42 million tons of crude oil. Large amounts of crude oil drifted ashore, leading to estimated deaths of 400,000 sea birds and 3,000 sea otters. Exxon, the company responsible for the accident, paid for the costs to clean up the oil and for damages to the fishing industry. However, as the U.S. Oil Contamination Law precludes that compensation for environmental degradation is also applicable in the case of oil spill accidents, both the Alaskan and federal governments demanded that Exxon paid compensation for damages to the ecosystem brought about by the oil spill. Litigation took place to determine the reparations.

In the trial, CVM was used as a calculation basis for the amount of damage to the ecosystem. First, respondents were informed of the conditions surrounding the *Valdez* oil spill. Next, a guard ship known as an "escort ship" was dispatched, and countermeasures to contain the accident were indicted. The, with the accident successfully contained, people were asked how much they would be willing to pay to preserve the Alaskan wildlife. In 1991, a questionnaire survey of a random lottery

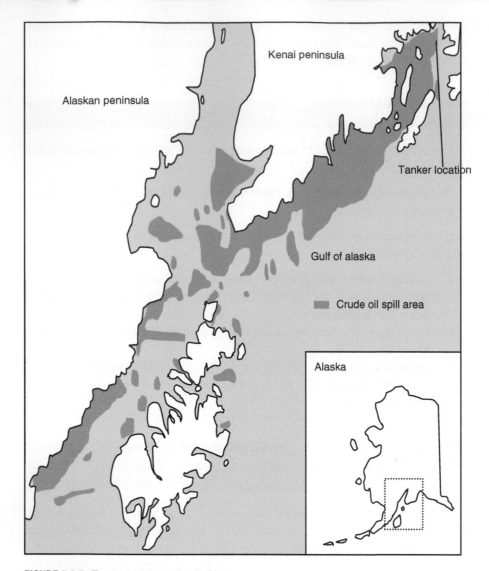

FIGURE 5.3.5 *Exxon Valdez* tanker incident

Source: Data collected from oil spill area public information center between March, 24 and May, 18, 1989.

sample of average American homes was performed, collecting valid responses from 1,043 households. Statistical analysis indicated that the marginal WTP per household was 30 dollars. By multiplying this value by the total 90 million households across the U.S., the $2.8 billion total cost was calculated. The trial proceeded based upon this damage compensation assessment, and in the end, it was determined that an additional 1.2 to 1.5 million yen would be paid in compensation for destruction of the ecosystem in out-of-court settlements. This is a real-life example of how CVM was used effectively in the court, leading it to garner attention across the globe.

Section 5.4: Cost-benefit analysis

What are cost-benefit analyses?

Previous sections examined environmental evaluation methods. This section takes a look at how those methods are practically applied. *Cost-benefit analyses* are used to evaluate public projects provided by the government. When people make decisions, they weigh the gains against the losses, or the advantages against the disadvantages, in order to bring about the greatest possible difference between gains and losses. In other words, the aim is to maximize the net benefits of choices thorough decision-making. Here, the words *gain* and *loss* can be replaced with the words *benefit* and *cost*, respectively.

The aim of cost-benefit analyses is to express benefits with monetary units and to support social decision-making. In this section, the cost-benefit analyses take into account the costs and benefits to the whole of society, so they are known as social cost-benefit analyses. When governments get involved in public projects, they must comprehensively evaluate the "before" and "after" social costs and social benefits of the project. During the examination, in order to determine how resources should be distributed among specified businesses, it is essential to have *a priori* evaluations. Moreover, in order to investigate the actual effect that the business had, ex-post estimates are necessary. Thus, comprehensive assessments that include both estimations are quintessential. Furthermore, cost-benefit analyses of the environment depict the necessary costs to implement environmental policy as well as the monetary amount of benefits gained from enforcing the environmental policy, and by comparing both, they indicate the efficiency of an environmental policy.

In market economics, whether or not investment plans are socially desirable depends upon whether revenue increases due to investment will surpass the costs of investing. If the market is functioning well, then resources and services that bring about efficient resource allocation will accompany a viable investment plan. Accordingly, investment decisions based on viability are socially desirable from perspectives of efficiency. However, if market failures arise, public components become the basis of intervention for people's investment plans. Especially with cost-benefit analysis of environmental policies, targeting such natural environmental aspects as air, water, and wildlife requires monetary assessments of environmental policy since market prices for the natural environment do not exist. The assessment methods for ascribing monetary unit value to the environment were explained in the first three sections of this chapter. Table 5.4.1 classifies cost-benefit analyses among adopted environmental policy precedents. Formats applicable to policy are benefit analyses that assess the benefit of environmental improvement, damage analyses that assess the damage costs of environmental deterioration, and comprehensive assessments that assess both the benefits and damages.

In U.S. *public works*, all plans undergo cost-benefit analyses, which must, in principle, be made publicly available. In Japan as well, cost-benefit analyses have become normalized, obtaining an important position among public policy evaluations. In April, 2002, the Administration Evaluation Method (Government Policies Evaluation

Table 5.4.1 Cost-benefit analyses incorporated in environmental policies

	Evaluation of public works	Evaluation of environmental regulations	Evaluation of natural resource damages	Environmental finance
Details	Use monetary unit measurements in evaluations of public goods and perform cost-benefit analyses of the business	Use monetary unit measurements in evaluations of the economic effects of environmental policies and perform cost-benefit analyses of regulatory policies	Evaluate the damage costs environmental losses brought about by oil spills and soil contamination; used by courts in determination of damage	Monetarily assess and compare the costs and results of a business or local government's environmental policies
Example	Evaluations of the effectiveness of: park maintenance, flood prevention, plumbing, economic effects of environmental policies	Evaluations of the effectiveness of: exhaust gas regulations, harmful substance regulations, water pollution substance regulations, diverse functionality of preserved farmlands	Damage assessment of the *Exxon Valdez* oil spill	Ministry of Environment's Environmental Finance Guidelines, life-cycle impact assessment method based on endpoint modeling (LIME), company environmental reports
International precedents	O	O	O	△
Domestic precedents	O	△	×	O

Note: O: Implemented, △: Partially implemented, ×: No applicable example

Act) officially obliged the government to perform policy assessments, emphasizing their importance. This led to an objective and strict adherence to policy assessments, resulting in appropriate policy reflections.

Furthermore, making information regarding policy assessment public, and investing in effective and efficient policy promotion, became a priority. In Japan, while it is true that cost-benefit analyses of public works are being performed, for fields outside of public works such as regulatory and tax systems, there remain many issues with the content of such areas.

Applications of regulatory policies

Unlike public works that are provided through tax money, the social costs of regulatory systems can be difficult to envision, leading to the adoption of inefficient policies. In particular, environmental and safety regulatory systems that could easily be drawn from emotional sentiments could have remarkable effects. Table 5.4.2 depicts the estimations of the necessary social cost of saving individual lives in the U.S. seat belt and airbag regulations as amounting to roughly $10,000, allowing one to grasp the cheapest way to save a person's life. In contrast to this, asbestos use regulations are estimated to amount to social costs worth around $300,000,000. Moreover, regulation related to formaldehyde and timber preservation chemicals are extremely difficult to eliminate and survey, so they cost enormous sums of roughly $2,500,000,000 and $170,000,000,000, respectively. Because of the sentimental and other reasons related to regulations of substances like formaldehyde, it has come to be recognized that they will bring about enormous social costs.

Furthermore, other precedents include the benefit assessments among numerous regulatory policies based on the *Clean Air Act* and *Clean Water Act* implemented by the U.S. EPA. In 1997, the EPA publicly announced the cost-benefit analyses of environmental purification methods that took place between 1970–1990. This was done in order to provide an ex-post assessment of the costs and benefits that were yielded alongside the *Clean Air Act* in comparison to times when it was not in place. Among the benefits included effects that air pollution regulations had on the prevention of health problems.

Table 5.4.2 Social cost involved in saving a single human life

Regulation/Restriction	Social cost necessary to save one life
Seatbelt and airbag regulations	Roughly $100,000
Asbestos regulations	Roughly $300 million
Restrictions on exposure to formaldehyde at the workplace	Roughly $256.4 billion
Table of dangerous wastes among chemical substances for wood preservation	Roughly $16.9524 trillion

Source: Viscusi, W.K., J.K. Hakes, and A. Carlin. 1997. Measures of Mortality Risks, *Journal of Risk and Uncertainty*, 14(3), 213–233.

Applications of public sector investment

When the public sector independently implements investment plans, the private sector resources only decrease by amounts equivalent to what is passed on to the public sector, and consumption and private investment are sacrificed. Accordingly, beyond the public investment and benefit sacrifice brought about by consumption and private investment, also with regards to determining public investment, the costs attributed to private investment, otherwise known as the public investment opportunity costs, must also be properly considered. In other words, even in cases where market failures are present, it is important to make realistic indications about superior, efficient choices rather than seek government intervention. Because of this, cost-benefit analyses are sought.

While the benefits to public investments are multifarious, indicating the monetary unit value of all possible aspects is evaluated and overall social benefits are calculated from this. The cost-benefit analysis is the comparison between the social benefits derived here and the investment costs. When public investments are assessed, the ratio of benefits to costs are not considered on the individual level, but the ratio of costs to benefits to society must be reflected. There has often been conflict between urban development and natural preservation objectives that arises among public works throughout Japan. The Nagara estuary weir gate shutdown of 1995, the closing of the Isahaya Bay tideland reclamation embankment project in 1997, the cancellation of the Fujimae tideland reclamation project in 1999, the local resident referendum regarding the Yoshino River gates in 2000, and other opposition movements gained profound momentum throughout Japan.

When a dam development business, for example, is assessed, it could be expected to assist in preventing flooding or providing electricity. There are also recreation effects, such as the ability to enjoy sightseeing after camps are constructed around dam reservoir areas. However, in recent times, concern for environmental issues continues to rise and more people believe that the natural environment should be preserved. Therefore, the ecosystem destruction damanges and flood prevention and electricity benefits, as well as the inherent embankment, water source maintenance, and resident displacement reparation costs, must all be included in business cost calculations when the dam is being constructed.

Next, if road construction were financed by public investment, the alleviation of traffic congestion that leads to saved time by users of the road, as well as the decreased travel costs, abated accident numbers, and reduced noise pollution, and other such benefits are calculated and ascribed monetary unit values and furthermore compared with the investment costs of the road construction. These benefits can also be the drawbacks of noise and air pollution that are brought about directly by road transportation, and in such cases are not social benefits but become social costs.

Costs are determined by the scope of the usage, reparation, construction, and other pertinent costs. Benefits are measured through assessments of people and commodity time values, loss profits, medical fees, mental damage of personal damage costs, and risks of disaster businesses, as well as environmental quality value. Especially, the influence that occurs with relation to the environment that is assessed monetarily has many extremely difficult issues. Here, the substitution law, the hedonic law, CVM (contingent valuation method), travel cost method, and other measurement methods are used to make calculations. Furthermore, with regards to CO_2, there are emissions trading markets that are put in place, so there are methods for setting based on emissions trading prices. Table 5.4.3 public works projects in Japan that utilize cost-benefit analyses. In this way, the business content that spans multiple areas is divided and the business benefits are classified, and as it is necessary to measure the benefits of each business, the assessment procedures are determined.

Ways to think about Pareto optimality

Within economics, *Pareto optimality* is commonly referred to in discussions of efficiency. Economic efficiency in terms of Pareto optimality implies that no single entity's utility can increase while someone else's decreases. That is, social welfare is improved to the extent that one person's utility increases without decreasing another person's utility. Accordingly, this is the foundation upon which the concept of Pareto improvements is based. As an example, consider a given condition *A* where a dam is to be constructed. Here, an individual's current conditions make *A* desirable, *B* are the benefits, and *C* are the costs, so the net benefits are (*B* − *C*), which amount to an overall

Table 5.4.3 Public works project assessments

Business description	
Roads and railways	Road projects
	Farm roads
	Roads (harbors and fishing ports)
	Railway projects
Port maintenance	Airport projects
	Harbor projects
	Fishing port projects
National territory and land preservation	Rivers and dams
	Erosion control projects
	Forestry conservation projects
	Shore projects
Sewage and waterways	Waterway projects
	Sewage projects
	Agricultural village drainage projects
	Fishery village drainage projects
Housing and cities	Urban area redevelopment projects
	Land maintenance/readjustment projects
Farm villages and forests	Agricultural village maintenance projects
	Forest environment and preservation projects
Public parks	Maintenance of city parks
	Maintenance of natural parks
	Harbor green tracts

Source: Kuriyama, Koichi. "Public Works and Environmental Assessments: The Role of Cost-Benefit Analysis in Environmental Assessments", *Japanese Annual Report on Environmental Economics & Policy*, 2003, pp. 55–67.

gain. If condition A were desirable for everyone, then the social decision-making would drive condition A into existence. Additionally, if many people want A, and the remaining people do not care whether or not A is implemented, such a scenario would also lead to A. However, there are many people who want A as well as many people who do not want A. Here, it is impossible to realistically determine a policy that improves conditions for everyone. In similar public investment project assessments, there are very few cases in which Pareto optimality is a valid standard. Instead, many experience welfare improvements while some experience losses in utility. For example, consider public investment in local railways. When this kind of investment occurs, the utility for those who live along the railways increases, but the land values for those who live in other areas goes down, so the latter see a decrease in their utility standard.

Just as the previously mentioned reasoning shows, an outline for the efficiency of cost-benefit analyses applied to public investments is, by default, different from Pareto optimality. With more simplified and realistically efficient guidelines, all benefits can be expressed with monetary units, and regardless of who gets their benefits returned, the monetary unit values are added up to a net total. However, this efficiency is problematic in that it assumes that 1 yen is valued the same by the upper class as by the lower class.

Therefore, for the whole of society, when the question is whether or not conditions being implemented are ideal for society, comparisons of the degree to which the utility of individuals' changes must be decided and standardized. When the monetary units are simply added together and agreement is easily conferred as a guideline, what could be called the "efficiency improvement" of the policy is "the people who have gains from a policy compensate those who lose out from a policy, and both parties gain an advantageous position." For example, when commuter railways are constructed, utility for those who live along the lines accrues, while utility for those who live elsewhere decreases. However, those who live along the railways could compensate others for their loss and still witness increases in utility compared to previous levels. This scenario would satisfy the aforementioned conditions.

Consider a decision-making problem among four individuals. Due to a transition to the new conditions, two people yield net profits, while on the other hand two people experience large losses. The two people with net profits each gain +10, while the two who lose are each out by −6. The total net benefits are ($2 \times 10 - 2 \times 6 = 8$). Provisionally, the two who gain a net profit compensate the two who experienced losses, and overall, everyone's benefits go up, amounting too net profits in the end. For this to play out, the two who earned net profits allocate 7 points to compensate those who incurred losses. Those who originally had a net profit still gain ($10 - 7 = 3$) while those who initially suffered losses also realize net gain ($-6 + 7 = 1$). If this kind of compensation is practical, then Pareto improvements can be achieved in reality. For, Pareto improvement is a scenario in which, regardless of how much compensation is provided, everyone's utility is raised through it.

Here, B_i is individual i's benefit, C_i is individual i's cost, and to the extent that the net benefit ($B - C$) is great, it becomes a socially desirable condition for N people. This all-individual decision-making that comprises society is dealt with by calculating the total value of each individual's net benefit. Then, if this value is positive, then that plan (the policy or business project) is justified. In other words, it is based on the equation:

$$[(B_1 - C_1) + (B_2 - C_2) + \ldots + (B_N - C_N)] > 0$$

One substitution method is adopted if the plan's net benefit is greater than zero. Furthermore, if there are multiple substitution proposals, the one that maximizes the net benefits is chosen. Then, if the net benefits that make net benefits positive do not exist, then it is said that there is definitely no special policy to improve current conditions, and longstanding conditions continue on as they are.

SUMMARY

The goal of cost-benefit analyses is to express the monetary value of benefits in order to assist social decision-making. The benefits and costs of public investments are not yielded at one point in time only. Generally, they are yielded over the long term, a period of many years. Because of this, the discounted present value discounted from the future benefit and costs with the discount rate is useful in making comparisons.

REVIEW PROBLEMS

1. Explain the benefits that ought to be yielded when public investments are used for road construction.
2. Explain why efficiency guidelines for cost-benefit analyses applied to public investments must necessarily differ from Pareto optimality.
3. Compare the two projects in the **Learning point**. Calculate the two scenarios in which the discount rates are 4% and 10%.

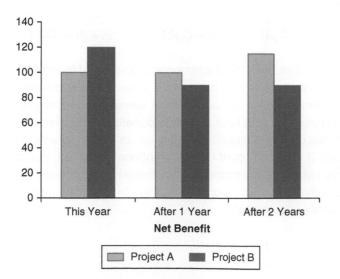

Net Benefit

LEARNING POINT: DISCOUNT RATES IN COST-BENEFIT ANALYSES

The benefits and costs of public investment do not arise at one single point in time, but rather, they are produced perpetually over the long term. There are benefits and costs that arise continuously over large, long-term projects, those that exert influence on the spot and then immediately disappear in short-term projects, as well as those that are produced after a certain amount of time has passed. Thus, methods that line up with the cost and benefits of each period of time are quintessential.

In order to compare the costs and benefits at different periods of time, perpetually, a *discount rate* is applied to compare future values with the discounted present value. The reason discounting is essential is because, even after the many people have consumed various goods, there are still preferences among present consumption choices. Furthermore, even directly after consumption, consumption opportunities could be abandoned when provisions are made to wait until a later time. This is known as people's time preferences. Here, if B_i is the benefit at time i, C_i is the cost at point i, and r is the discount rate, then the costs and benefits yielded after one year are equivalent to $(1 + r)^1$, and the costs and benefits yielded after some t years, expressed by $(1 + r)^t$, is used to calculate the present discount value.

Now, consider a three-year project. Assume a 5% discount rate and that yearly net benefits amount to 100,000 yen. In this case, the present discounted value equates to:

100,000 yen + (100,000 yen)/1 + 0.05 + (100,000 yen)/$(1 + 0.05)^2$ = 100,000 yen + 95,200 yen + 90,700 yen = 285,900 yen

Expressed simply, the present discounted value of net benefits from time 0 to T can be depicted by:

$$(B_0 - C_0) + (B_1 - C_1)/(1 + r) + (B_2 - C_2)/(1 + r)^2 + \ldots + (B_T - C_T)/(1 + r)^T$$

If this value is positive, then the implementing the project is justified.

The discount rate used in such a case is known as the discount rate of public investment. Determining the discount rate for sewage and many public facilities that are to be used over the long term and yield benefits has proven to be considerably difficult. The discount rate standard largely exists in the present discount value of social benefits. If a low discount rate is intentionally applied, the present discount value of social benefits is maintained at an excessively high level, and public investment consistently takes place. Conversely, if too high a discount rate is used, public investment barely occurs at all.

Business and environmental problems

Chapter overview

This chapter focuses on corporate environmental practices. Since social concern for environmental problems has elevated, voices calling for businesses to have environmental policies have grown louder. Namely, there is a greater demand for "environmental practices," or business practices which give due consideration to environmental issues. Especially in recent years, emphasis has been placed on expanding business considerations beyond the traditional economic and legal responsibilities that businesses have been obliged to abide by to stakeholder considerations. For, it is the stakeholders who have vested interests in the wellbeing of businesses. Finally, the importance of corporate social responsibility (CSR) has come into the limelight. This chapter introduces the nature of various environmental risks that businesses must address. Furthermore, it explains various methods for businesses to accomplish this.

Chapter content

Section 6.1—This section introduces the means for businesses to efficiently execute environmental policies. First, a multi-step analysis of business processes, from raw material extraction to waste disposal, is employed to shed light on the many ways that businesses may cause environmental harm through their operations. Then, the costs and effects of environmental accounting and related policies are considered.

Section 6.2—This section provides an in-depth explanation of corporate social responsibility (CSR). Beginning with the background conditions that have led to the creation of CSR, this section defines the term and follows with depictions of its long-term benefits with regards to future social welfare fluctuations. The section closes with investigations of socially responsible investments and environmentally conscious finance.

Section 6.3—Since businesses have a hand in environmental pollution and other risks, it is essential to assess environmental risks and have ready-to-implement countermeasures before contamination accidents take place. This section outlines environmental risk fundamentals and explains the environmental risk assessment methods.

Section 6.4—The final section of this chapter depicts how businesses influence the ecosystem and biodiversity. Since preserving wildlife requires large sums of funding, businesses and other large, private capital holders are often expected to finance projects to preserve biological diversity. The various initiatives for preserving biological diversity are laid out here.

Section 6.1: Businesses and environmental policies

Environmental management and environmental management systems

Businesses ought to play an important role in tackling environmental challenges. Consider climate change issues as an example. In 2009, Japan's CO_2 emissions were 1,209,000,000 tons, and of this amount, emissions from the manufacturing sector comprised 35.2%, those from the office sector comprised 19%, and those from the shipping sector comprised 20%. In other words, the previously mentioned business sectors largely surpassed the 14% that the household sector emissions made up of total national amounts. Accordingly, reducing business emissions is quintessential to preventing climate change.

As social concern for environmental issues continues to rise, so too do demands for businesses to adopt environmentally-friendly practices. And, *environmental management*, or business practices that give due consideration to environmental issues, have gained prominence. Up until now, the primary concern of businesses was to pursue profits, and since environmental policies were costly and did not directly tie into profit earnings, environmental policies had been thought of as at odds with business management. However, nowadays, businesses can no longer ignore environmental issues as part of their social responsibilities. Figure 6.1.1 depicts the proportions of businesses that have been drawing up business policies related to the environment from the year 1991 through 2005. While in the year 1991, a mere 30% of large enterprises took multiple environmental issues onboard through management strategies. Beyond this information, by 2010, over 80% of large businesses implemented these methods, alluding to the fact that the heads of corporate entities have made environmental policies some of the most important considerations among management principles.

The procedures through which these environmental practice fundamentals are reflected in actual business activity are known as *environmental management systems* (see Figure 6.1.2). The International Standardization Organization (ISO), which sets

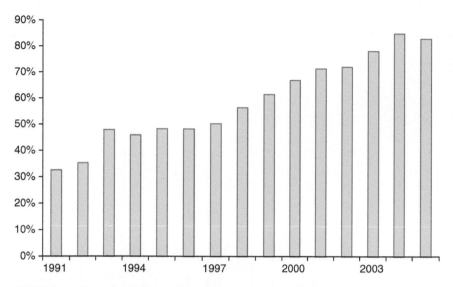

FIGURE 6.1.1 Proportion of firms with environmental management policies

Source: "Survey of Environmentally-friendly Business Activities" published by the Environmental Economics Section, Environmental Policy Bureau, Ministry of Environment, Japan [Japanese]

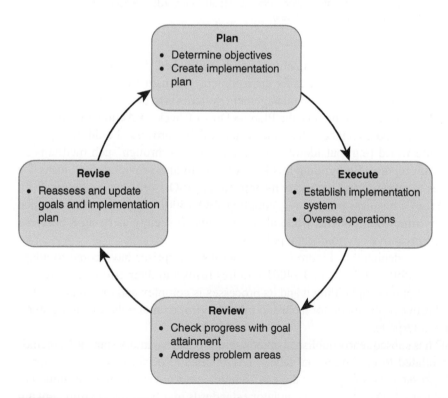

FIGURE 6.1.2 Environmental management system

Source: Created by author with information from the Ministry of Environment, Japan

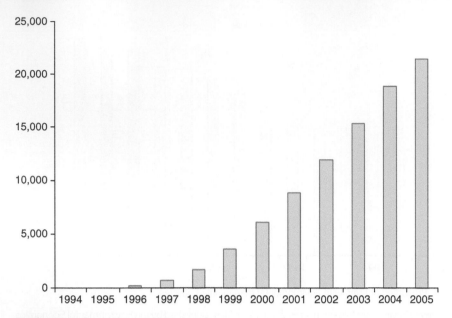

FIGURE 6.1.3 ISO 14001 registration trends

Source: Created by authors with information from the Japan Standards Association (Environmental Management Specifications Affairs Committee)

international regulations for manufactured goods, published ISO 14004 in 1996 as a guideline for international standards for environmental management.

The first aspect of environmental management systems is to determine environmental policy objectives. Based on this, the Plan → Do → Check → Action, or the PDCA cycle, is derived and executed. Then, the results of this process should be reported to the outside world (e.g. outside of the company itself) through such mediums as environmental reports. By building this kind of system and providing the means for third parties to research and recognize its application, ISO 14001 can be successfully adopted. From a business perspective, adopting ISO 14001 allows firms to advance efficient environmental policies while also appealing to society as being a business that upholds proactive environmental policies.

Therefore, as depicted in Figure 6.1.3, numerous enterprises have come to adopt ISO 14001. Unfortunately, as ISO 14001 requires firms and their efforts to be recognized by third parties, upholding it and its processes is complex and extremely costly. This could prove problematic for inclusion in the environmental policies among small to medium enterprises.

The ISO has subsequently published other international regulatory standards, including those related to environmental labels, environmental performance assessments, lifecycle assessments, designs for the environment, and environmental communication. All of these ISO international regulatory standards involving the environment are included in what is known as the ISO 14000 series.

Lifecycle assessment (LCA)

When businesses are determining environmental policies, or when consumers aim to select products that are environmentally friendly, it is important to understand which technologies and products place what amount of strain on the environment. Moreover, when contemplating a product's environmental impact, one must duly consider every stage in that product's existence, from raw material procurement for its creation to its disposal as waste. For example, new refrigerator models may be developed in such ways that they conserve more and more energy than other products up until now, and they may even emit negligible CO_2 over the life of their use. However, their manufacturing processes could place a greater burden on the environment than those of other products up until now. In this scenario, the decision of whether or not to choose an outdated model or new model is rooted in different judgment, whether one is considering only the manufacturing stage of the good or if one is considering everything through the usage stage.

In this way, in order to appropriately grasp a product's environmental impact, one must look into all of a product's stages, from raw materials acquisition → manufacturing → distribution → use → disposal stages, or the lifecycle of the product, which includes the environmental burden yielded at each stage of the product's existence. The method of investigating these various environmental burdens that a product yields "from the cradle to the grave" is known as *lifecycle assessment*, or *LCA*.

In order to execute LCA, the "piling up method" in which all of the environmental burdens yielded from individual processes are culminated together, as well as the "inter-industry relations analysis method" in which environmental burden output levels of emissions from each industrial sector are used based on an inter-industry relations chart, are both utilized.

By making use of LCA, the extent to which greenhouse gases, air contamination substances, water contamination substances, toxic substances, waste matter, and all

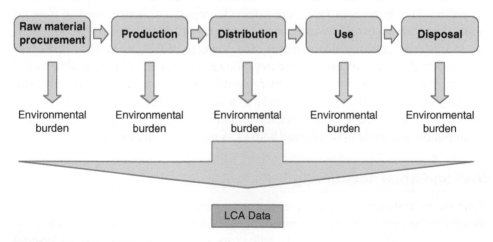

FIGURE 6.1.4 LCA components

other burdens are placed on the environment throughout a product's lifecycle can be understood. However, even through using LCA, there are cases where judgments are split, alluding to the inescapable nature of some environmental burdening. For example, consider the case in which collected plastic bottles are to be renewed into new plastic bottles. If used plastic bottles were to be buried as they are, they would become waste matter. On the other hand, if the plastic bottles were to be renewed, while this action would prevent the creation of waste matter, CO_2 would be emitted at the collection and restoration processing stages, as immense amounts of energy would be required. In other words, either process creates waste matter, and either process contributes to climate change. Since waste matter issues and climate change issues cannot be compared by any simple means, the only judgment that can take place is which environmental burden is preferable.

Accordingly, there has been progress with the development of the LCA "integrated index," in which various environmental burdens are integrated into a single index. In order to integrate climate change issues, waste matter issues, health damages, and other various environmental aspects, each issue must be ascribed a numerical weight. As it is essential to place a value judgment on whether climate change or waste matter issues is more important, this numerical weighting must not be based on measuring the physical material units of environmental burdens alone. One method is base weighted values of environmental issues on expert opinions. For example, in the *Eco Indicator 99* published by the Netherlands, LCA experts prioritized human health, ecosystem health, and resources as the three most important categories, and integration was performed based on these priority weightings.

Another integration method involves a money assessment. If all the environmental burdens were assessed using monetary units, it would be possible to summarize a single index. In Europe, the EPS (environmental priority strategy, Sweden) and ExternE (energy externalities, European Commission) use CVM (contingent valuation method) to make monetary assessments of environmental burdens, leading to a total sum of damage costs of said environmental burdens that subsequently lead to integration. Furthermore, in Japan, the Lifecycle impact assessment method base on endpoint modeling (LIME) is under development, but LIME utilizes conjoint analysis to integrate environmental burdens. The impacts of CO_2, SO_2, NO_x, and other environmental burdens on the four preservation targets, namely human health, public property, biological diversity, and primary production output, are arranged, and with monetary evaluations, they are integrated into a singular index. Although the development of integrated indices are still in the research phase, they have garnered attention as compilations capable of combining both LCA and economic evaluations.

Environmental accounting

While an understanding of the immense costs incurred by businesses that undertake environmental policies is indeed important, in order to implement highly effective environmental policies with relatively little costs, one must appropriately understand the costs and effects of said policies. The aim of *environmental accounting* is for businesses or governments to recognize the costs and effects of environmental policies by comparing the two, and to subsequently and voluntarily convey that information to outside parties in an easy-to-understand way. In environmental accounting, business managers utilize "internal environmental accounting" to investigate their own environmental policies within the company, as well as "external environmental accounting" where their aim is to share information with parties outside of the company.

In Japan, the former Ministry of the Environment of 1999 presented the *Guidelines for Understanding and Publication of Environmental Conservation Costs* which set the precedent for the adoption of environmental accounting. The 1999 guidelines only aims at understanding the costs of environmental costs, but within the guidelines published in 2000, assessments of both cost and effects of environmental policies were targeted. Afterwards came the 2002 and 2005 revised editions, which led to the current *Environmental Accounting Guidelines*. The guidelines of the modern Ministry of the Environment are similar external environmental accounting in nature, with the aim of disclosing information to outside companies. Through the Ministry of the Environment's guidelines, progress continues to be made with standardizing environmental accounting, and in recent years, close to 800 companies have employed environmental accounting (See Figure 6.1.5).

Within the environmental preservation costs statistics of environmental accounting guidelines, internal business costs of environmental policies that arise among factories are compiled, and the upstream and downstream costs of environmental policies, from raw material provision to post-sale stages, are compiled. Management activity costs are those related to environmental management. Research and development costs are compiled based on the expenses related to environmental policy R&D, and social activity costs are classified as the contributions to environmental conservation organizations. Finally, when an enterprise makes damage compensations for soil contamination and other environmental pollution that it may have caused, such expenses are categorized as environmental damage costs.

On the other hand, the energy and resources invested into business activity, as well as the environmental burden yielded by business activity, are calculated in physical

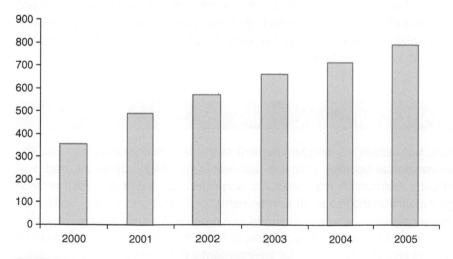

FIGURE 6.1.5 Number of firms using environmental accounting

units with regards to the effect of environmental policies. By comparing these with the environmental policy expenses, it becomes possible for an enterprise to understand what sorts of expenses would be involved in decreasing environmental burdens. However, as the effects of environmental policies are measured in physical units, they cannot be used to judge whether the policy led to a budget surplus or deficit. Therefore, some have proposed that the economic effects of environmental policies should be indicated. For example, classifying the profits earned through selling recycled products can be labeled as revenues, while the expenses curtailed through energy and resource conservation can be classified under the expense reduction category.

However, these economic effects will at most only be concerned with the private effects that are yielded within the business itself and they do not include the externality effects that are produced outside of the business. For example, risks of damages induced by flooding and droughts, contributions to reducing risks of large-scale damages like wildlife extinction, and other effects of climate change that are yielded outside of businesses themselves are not included as in the effects of current environmental accounting guidelines when climate change policy is implemented. The only result of climate change policy that can statistically be quantified is how much fuel cost was economized through energy conservation policy. Of course, there are many precedent cases in which the goal of companies implementing climate change policies was not only to reduce on fuel costs, but also to actually undertake social responsibility. However, since assessing the externality effects is no simple feat, they are not included as a sum of money among environmental accounting guideline calculations. Thus, there are many cases where publicly announced environmental accounting includes costs that surpass the value of positive policy effects, making the policy a source of deficits.

This means that, while it is so that external environmental accounting has proliferated since the Ministry of Environment presented its guidelines, modern guidelines do not appropriately assess the externality effects of environmental policies. Indeed, the problem remains that the effects of such policies are largely under-assessed. In order to evaluate the external effects in the present day, CVM and other environmental evaluation methods are essential, and since making such evaluations requires expert knowledge, it is difficult for many firms to perform these assessments independently. Thus, developing methods for easily assessing external effects remains a pertinent topic to be addressed in the future.

SUMMARY

Nowadays, businesses rely on management systems to successfully execute efficient environmental policies. Lifecycle assessments, which outline all stages of environmental interactions from resource acquisition to waste disposal, provide thorough insight into the impact of environmental policies. Furthermore, while environmental accounting has rapidly emerged as a means to compare and contrast the costs and effects of environmental policies, at its current stage, it is undermined by a tendency to excessively undervalue environmental policy impacts.

REVIEW PROBLEMS

1. What are the merits and drawbacks of adopting ISO 14001?
2. Non-physical, monetary units are used to depict environmental burdens in LCA. Explain the drawbacks as well as the importance of using monetary unit assessments.
3. Examine the environmental accounting published in the environmental report of a company that you know. Then, explain the special features and problematic aspects of the company's environmental policy.

LEARNING POINT: LCA PRECEDENT: ERASABLE PRINTING TONER

In order to take a realistic look at the LCA analyses method, consider a pertinent example involving printer toner. Until recently, printing toner could not be reused once a print was made. More recently, with the development of erasable toner, printed characters could be erased with eraser machines, allowing the printed-paper to be reused. However, eraser machines could inflict different sort of environmental harm. Therefore, the extent to which negative environmental impacts are reduced by reusing paper must be compared with the potential harm produced by eraser machines and must be calculated in order to judge the merits of using erasable vs. conventional printer toners. Figure 6.1.6 depicts the results of an analysis of erasable toner through the LIME framework. Accordingly, the resource and ecosystem conservation effects brought about by reusing paper is greater than the increased environmental damages brought about by implementing eraser machines. Thus, the erasable toner has been determined as more environmentally friendly than the toners used beforehand.

Section 6.2: Social responsibility demanded from businesses

Corporate social responsibility (CSR)

In recent years, thoughts about *corporate social responsibility (CSR),* which expands the traditional economic and legal responsibilities of businesses up until now to include considerations of company stakeholders, have become prominent. Among real business practices, in addition to the ethical and moral information included in environmental reports up until now, the publications of CSR reports and sustainability reports are rapidly increasing (see Figure 6.2.1). For example, of the top 250 *Fortune Global 500* companies, 45% in 2002, and 52% in 2005 independently published CSR reports, denoting a large-scale increasing trend. Furthermore, Japan is the country with the highest percentage of companies that create CSR reports.

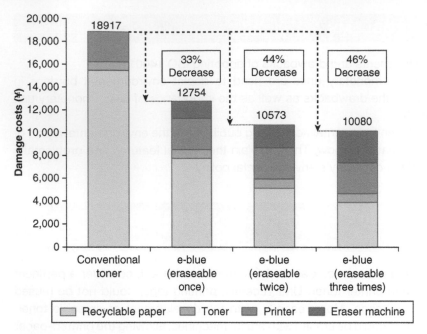

FIGURE 6.1.6 LCA analysis of erasable toner

Source: Itsubo, N., Inaba, A. (eds), 2005, "Life Cycle Impact Assessment LIME: LCA, Environmental Accounting, Assessment Method and Database for Environmental Efficiency", Maruzen [Japanese]

Many wonder why CSR is drawing so much attention. While many reasons for this exist in the context of the modern, globalizing, information age, the simple answer is that companies can no longer operate with the sole economic aim of generating profits, and must also incorporate decision-making procedures that have comprehensively reflected economic, environmental, and social impacts. Moreover, they must take great care to recognize the importance on their relationships with various stakeholders (stock owners, employees, consumers, the environment, the community, and others who have influence on the company) while pursuing structural changes towards an organization that contributes to the realization of sustainable society (see Figure 6.2.2). On the other hand, from a business perspective, one could say that CSR could become a source of competitive superiority in markets over the long-term. Provisionally, as suggested by the latter point, if CSR activity promotes conditions that raise the company's profits, then the company would have no issue with adopting it.

This section explains what enterprises should consider for CSR, and furthermore, whether or not a business is capable of or should engage in CSR. In order to do that, one must first gain a firm understanding of CSR and its components.

What is CSR?

The presence of a business in a region or locale is directly related to the total amount of goods and services that it produces and supplies to nearby consumers. Moreover,

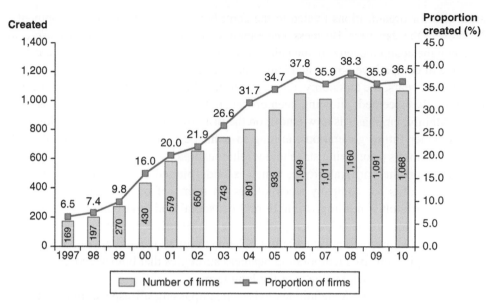

FIGURE 6.2.1 Trends in the number of firms that produce environmental reports

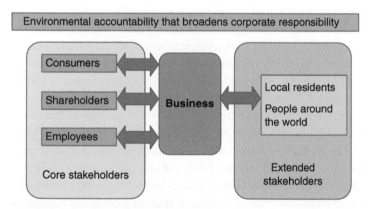

FIGURE 6.2.2 Stakeholder expansion

business establishments tend to impact employment rates, household makeup, education systems, environmental quality, welfare systems, and a host of other societal aspects of local or greater communities. As said impacts are far from negligible, the purpose, value perspectives, and other directing principles of a firm must adequately reflect the needs of the times, stakeholder intents and benefits, and other interrelationships between the firm and its host community.

Due to the inherent complexity of said considerations, definitions of CSR have traditionally been derived from broad perspectives of what it ought to consist of. For instance, the Organization for Economic Cooperation and Development (OECD) Multinational Corporation Guidelines define CSR as "a contribution to economic, social, and environmental development, aiming to implement sustainable development." The Global Reporting Initiative (GRI) Guidelines define it as "declarations related to visions and strategies for organizations related to the contributions to sustainable development." Moreover, the Japanese Business Federation views it as being "comprehensive to grasping the economic, environmental, and social aspects of business activity, creating a source of competitive power, and improving business worth."

In recent years, the autonomy of every individual company and industry has come to be highly esteemed, allowing keen observers to recognize a convergence among ideas. Mainly, emphasis is now placed on the principles of sustainable development that consider future generations, aiming to enhance corporate value through such CSR behavior as prioritizing stakeholder relationships. Nowadays, many companies are taking on the corporate form, which is underpinned by a motivating priority to secure profits for company stockholders. In terms of business management, it would be difficult if not impossible to pull off social and environmental initiatives should stockholders perceive that profits were not of prime importance to business overseers. This means that by default, CSR must aim above and beyond traditional business aspirations by incorporating business strategies that both secure profits and improve corporate social value over the long term.

A broad array of CSR definitions exist, encompassing all aspects related to the impacts that businesses have on the environment and society. In this book, CSR is defined as "the rules beyond legal regulations that are voluntarily upheld by companies which improve *the environmental and social performance* of said companies." In other words, in addition to a company's legal compliance, environmental conservation, and engagement in human rights issues, its corporate responsibilities also include increasing long-term interests. And, all of these aspects add to the business enterprise's value as a member of society. By thinking in this way, it is possible to eliminate ambiguity and have fruitful discourse based on a clear definition.

Beyond voluntarily adhered-to rules that extend past established legal regulations, there are two types of CSR. One is tied into profit making while the other is not. CSR that undermines profits can be seen in cases where the shareholders act as greater stakeholders and the firm aims for social responsibility outside of monetary revenues. On the other hand, profit-inducing CSR is considered to bolster corporate value and profits over the long term, even in cases where increased operating costs and decreased profits are experienced in the short term.

Cases connected to long-term profit increases

The following five points are thought of as realistic conditions in which businesses perform CSR:

1. When a business' environmental friendly products can be set aside as distinguished,
2. When skilled labor can be attracted,
3. When the future risk of a firm can be reduced,
4. When good relations can be established with the government and local community, and
5. When the regulatory framework making process can be influenced and business conditions can be made comparatively more costly for a firm's competitors.

The subsequent section describes a rather concrete example of point number one, in which a product is distinguished from others.

As they directly translate into profits, supplying products that yield rather large environmental benefits, creating products through relatively inexpensive environmental policies, and finally, producing products and services with low environmental burden, are all prime examples of conditions tied into profit increases. In these cases, in order to successfully distinguish environmentally friendly products, the following three points must be satisfied:

1. The willingness of consumers to purchase environmental quality must be recognized, or otherwise, created.
2. Concrete information related to a product's environmental friendliness, as well as other characteristics, must be provided with certainty.
3. Innovation that cannot be replicated by competitors must be carried out.

The influence on social welfare

The final question is "Should businesses perform CSR?" While there are instances where CSR is defined as something that "aims for the realization of sustainable development," does CSR activity actually improve social welfare? If CSR led to declines in social welfare, would it become less appropriate for us to purport that businesses should engage in it?

About 70% of today's multinational business CEOs believe that CSR is indispensable to increasing their company revenues. Accordingly, companies estimate CSR to be a revenue-increasing activity. However, since financing is required for CSR activities, in order for businesses to loosen future regulatory levels, the result of CSR activities that is strategically performed towards government is that when regulatory levels are curtailed, there is, conversely, potential for social welfare to decrease.

In the same way, since insufficient regulations are only ever actually implemented (if regulations are put in place at all), CSR possesses great potential to increase social welfare. In the case of Japan, the adoption of sufficient regulations has primarily been hindered by the variability and social costs inherent among regulatory processes and the financial burdens placed on businesses in the form of environmental taxes. And, while the Kyoto Protocol called on developed countries to decrease emissions of six greenhouse gases compared with 1990 emissions levels over the 2008–2012 timeframe, the U.S. refused to ratify, severely undermining the momentum and efficacy of the Protocol's aims.

Therefore, rather than entrusting the management of environmental burden levels to governments, many cite the importance of alternatives methods which entrust companies to act voluntarily in CSR undertakings. A range of such voluntary participation programs, from pact agreements to information publications have gained prominence due to their usefulness in CSR aspirations. One example is the 313th article of the U.S. Emergency Planning and Community Right to Know Act (EPCRA) implementing the Toxic Release Inventory (TRI) system in 1986, which serves to emphasize the importance of information disclosure. This led to large-scale toxic chemical emission reductions throughout the U.S. Aiming to achieve similar ends in Japan, the Pollutant Release and Transfer Register (PRTR) Act was officially announced and instituted in July, 1999. This produced a shift in chemical substance risk management from the conventional direct, mandatory processes to discretionary business practices.

As another example, ten large companies in the U.S. provided insight for achieving CO_2 and other greenhouse gas reduction targets by up to 10% over the next 10 years in order to avoid climate change. Many policy principles are published within these reports, including holding responsibility for climate change from a global point of reference, creating incentives for technological innovations, and becoming environmentally efficient.

Compared to instances where regulatory systems are insufficiently or not at all implemented, through reliance on CSR activities, social welfare can be increased via voluntarily decreasing toxic chemical emissions, voluntarily decreasing CO_2 emissions, and other similar measures.

Socially responsible investment (SRI) and environmentally conscious finance

Next, we introduce methods for social responsibility through finance. For the finance world, separated as it is from the manufacturing industry, CSR initiatives center on *socially responsible investments (SRI)* that fulfill responsibilities to society through provision of capital resources. In recent years, as an evaluation standard for stock investments made in businesses, which are CSR activity-centered the SRI has grown in popularity. In addition to serving as a financial indicator, SRI up until now, has become more concerned with stability creation, societal contributions, environmental considerations, legal compliance, employment customs, respect for human rights, consumer issues, and other social and ethical standards. Accordingly, these standards have become the bases of evaluation, and companies that possess them typically receive investments. Furthermore, there is both social justice and regional contributions in SRI, and it also means that financial provision is a goal of the exercise of stockholder rights. Stocks are the standard investment means, but overseas bonds and other non-stock items are also common

Just as there is the question of whether or not CSR becomes a source of business profits, the finance world is interested in whether or not SRI funds can be anticipated to bring about higher returns on investment than normal funds. SRI funds are used to avoid recurring downturn shocks and to reduce *business risk* (including revenue changes over the long term). It is thereby considered to be the monetary amount needed to alleviate the risk of downturns. In other words, SRI can, to some degree, raise revenues and it is also capable of limiting revenue curtailments. To date, much analysis into whether SRI performance meets or exceeds that of other general funds persists, and the results of said analyses are not always clear enough to be reliable.

Environmentally conscious finance is yet another important endeavor in the finance sector. Wind power generation businesses that do not emit CO_2, sustainable resource waste-processing facilities, recycle electricity production businesses, and so on, all garner finance from environmentally conscious projects. Within environmentally conscious finance, the scope of the risk burdens for financial institutions becomes an issue. Originally, financial institutions bear additional risks depending upon the easing of lending conditions and favorability of interest rates. Moreover, it is imperative to run fiscally responsible businesses capable of generating information without sacrificing profits. In the case of interest rates and other privilege measures being put in place, through cost-benefit analysis, improvement of corporate value must be explained to shareholders and investors.

Profitability and realistic business management

Businesses that undertake CSR must rethink their decision-making processes, viewing standard business activities like production as inexorably linked to social and environmental issues. Such businesses must also strive to produce goods and services that lead to improvements in social and environmental quality. CSR is essentially a management strategy that is vital to synching corporate governance frameworks with efforts to maximize and distribute corporate value over the long term.

While engaged in voluntary procedures related to social and environmental problems, companies that face direct market competition are forced to ask themselves whether or not *it pays to be green.*

The contents of this section provide clear and concise definitions, allowing for discourse and explanations about CSR to follow. Yet in reality, there is a widely held belief that regardless of the presence or lack of CSR, each and every business is naturally obliged to adhere to the law, protect the environment, respect human rights, and uphold other aspects of civil code in common with all other members of society.

SUMMARY

In recent years, corporate social responsibility (CSR) has attracted a lot of attention. While there are diverse definitions for CSR, we defined it as "the rules that businesses adhere to above and beyond those required by law or set by regulatory systems which increase environmental and social performance." Through CSR activity, there are conditions that both increase and decrease gains for a business. In the same way, there are also conditions in which social welfare can increase or decrease.

REVIEW PROBLEMS

1. Define CSR.
2. Explain the kind of conditions in which it is practical for businesses to engage in CSR activities.
3. State the CSR activity conditions that are linked to increases in social welfare.

LEARNING POINT: THE UNDERPINNINGS OF CSR

As there is no general concept for what CSR is, it follows that it takes on many forms. This is because the national, territorial, business, timeframe, and other such contexts related to particular CSR cases all differ.

Milton Friedman developed a discourse based on the fundamental principles of corporations. He stressed the importance of stockholder rights, stating, "A company should endeavor only in so much as it maximizes profits, and it must not act in a manner that is at odds with this aim." CSR is charged with the task of acknowledging the priorities of maximizing profits and avoiding profit-diminishing endeavors.

CSR has recently made its way to the core of environmental management policies. Global environmental problems, being such large-scale issues, demand an overhaul of social economic system foundations. This would, as a matter of fact, have enormous impacts on business practices. Therefore, putting environmental economics into practice not only involves simply being socially responsible, but also places importance on exceptional business management strategies and requires analyses to be made from perspectives of business competitiveness.

However, regardless of the background conditions, what must always be borne in mind is that while firms do exist as social entities, they cannot continue to exist if they do not increase profits. The manner of responding to social demands is decided upon by each business upon its due consideration of how to adjust dividends to shareholders, prices of goods and services for consumers, employee wages, and other particulars. The benefit of CSR is that any third party can verify it. However, if it is not connected to specific profits, there will be no incentive to adopt it.

The discussions throughout this section focus on the relationships between environmental and social performance and profits and costs. However, for firms in Japan, the U.S., and Europe, the perception that the sole social responsibility is to aggrandize profits is changing greatly over time. And, there have even been instances where there is a clear, positive correlation between CSR pursuits and increased profits.

Section 6.3: Business and environmental risks

What are environmental risks?

Businesses take on a variety of risks, and among them, some are related to the environment. Environmental contamination that results from a mishap at a manufacturing plant, resulting in a cessation of plant operations and damage compensation to local residents, is a prime example. These types of risks that tie directly into environmental issues are known as *environmental risks*.

Depending upon the accident, environmental contamination could occur on such a large scale and inflict such serious harm to the surrounding environs that the continued existence of the responsible company could be threatened. Table 6.3.1 outlines large-scale contamination accidents, some of which harmed up to hundreds of thousands of people, others of which led to hundreds of millions of dollars in damage compensation.

The fact of the matter is, the chances of such large-scale contamination accidents occurring are so low that businesses owners tend to be ill prepared for their actual occurrence. The result is that if and when such catastrophes do take place, businesses are faced with paying exorbitant compensation fees. Thus, it is essential for policy to be in place before accidents occur so that businesses can appropriately understand the environmental risks they face. Environmental risk could be expressed by the equation "accident damages × accident occurrence likelihood." Accordingly, in order to grasp environmental risks, in addition to predicting the likelihood that an accident will occur, it is essential to accurately estimate how much damage would be inflicted if an accident were to actually occur.

Risk communication

If a given facility could potentially cause a contamination issue, even if an accident policy is in place, local residents who feel uneasy about such a happening could oppose or even prevent the facility's construction. For example, local residents who feared the danger of toxic chemicals contaminating groundwater could oppose the construction of the site. Such precedents, in which the opposing agendas grew worse and the construction eventually did not go through, have occurred in a number of different places. When a facility that poses some sort of contamination accident risk

Table 6.3.1 Severe contamination accidents

Year	Incident	Specifics
1976	Italy: Seveso chemical plant explosion	After a plant explosion, dioxins were released into surrounding areas, leading to health problems for over 220,000 people.
1984	India: Bhopal gas tragedy	Harmful gases emitted from a factory led to over 15,000 deaths and over 500,000 health injuries. Damage compensation amounted to $470 million.
1986	Chernobyl nuclear power plant disaster	A nuclear reactor explosion led to the widespread dissemination of radioactive materials, leading to 31 deaths and 135,000 displaced persons.
1989	*Exxon Valdez* oil spill	42 million liters of oil spilled into the Prince William Sound in Alaska, leading to the death of 400,000 sea birds and 3,000 sea otters. Fishery damage compensation amounted to $287 million, while compensation for ecosystem damages amounted to $1 billion.

is to be constructed, along with assessing the contamination accident risk ahead of time, it is essential to clearly articulate information related to environmental risk to local residents and to reflect local concerns in the planning of the facility. *Risk communication* is the collection of methods through which risk information is shared between business and local residents, leading to the eventual operational progress through mutual dialogue.

When conveying environmental risk to local residents, it is never easy to depict how large the risk actually is. Some residents who are uneasy about accidents may demand a "zero risk" assurance, alluding to the notion that no accident would occur at all. However, making any project "zero risk" involves such enormous costs that realizing it is impractical. On the other hand, if the probability of an accident is, say, 1,000,000 to 1, and this is conveyed to local residents, it would still be difficult to get people, who are yet unfamiliar with the probability of even everyday risks, to understand.

Psychological studies related to *risk awareness* have highlighted the fact that the severity of the risk recognized by common people differs from how large the risk of an accident really is. For example, consider the risk of death from nuclear power plant accidents and the risk of death from traffic accidents. Excluded are the yet unconfirmed effects of the Fukushima Daiichi Nuclear Power Station accident, the nuclear power plant mishaps across Japan that have led to deaths include the Tokaimura JCO criticality accident of 1999, in which two people passed away, and the Mihama Nuclear Power Station accident of 2004, in which five people passed away. While the number of deaths is few, the social concern for nuclear power stations has grown to extremely high levels, and newspapers and televisions report on them extensively. On the other hand, according to the National Police Agency, public transportation related deaths amounted to 4,612 people in the year 2011 alone, elucidate the fact that the risk of death from traffic accidents greatly exceeds the risk of death from nuclear power plant accidents. However, the media covers nuclear power plant accidents much more scrupulously than traffic accidents, and thus, few people actually believe that the risk of traffic accidents is more severe than the risk of nuclear power plant accidents.

Why is the risk that people perceive different from risks in reality? People's perception of risk is related to how scary they believe a matter to be, as well as how well they know the details pertinent to the matter. Figure 6.3.1 depicts the analysis of a questionnaire survey performed on U.S. citizens with regards to their risk perceptions. It is classified into two different axes: "Dreaded" vs. "Nonthreatening" and "Unknown" vs. "Familiar". Even through the probability of them occurring is low, nuclear power plant accidents and nuclear warfare are perceived to be "scary" because of the enormous damages they would produce were they to take place in reality. Caffeine and aspirin are viewed as "not scary" considering they are ordinarily ingested. On the other hand, cutting-edge technologies, such as genetic modification technologies, are perceived as "largely unfamiliar," and traffic accidents are viewed as "well known." Furthermore, since nuclear power plant accidents and genetic recombination technologies are perceived as scary and quite unfamiliar, general people view their risk with contempt. On the other hand, while traffic accidents are examples lost to home,

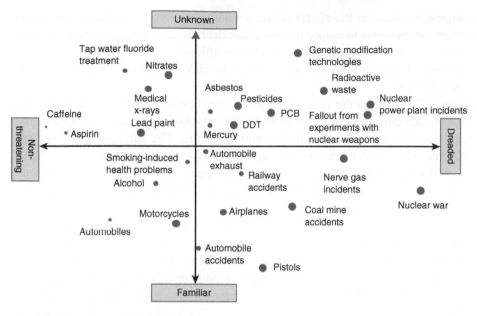

FIGURE 6.3.1 Common perceptions of risk

Source: Paul Slovic, Perception of Risk, *Science*, Vol.236, pp. 280–285, 1987

people feel as if they are quite familiar with their risk, rendering such risk to be considered less severe in nature.

Applying these concepts to environmental risks, toxic waste contaminating the groundwater and soil are not regularly occurring events, and ordinary people are not familiar with the toxic substances being used at factories, so such environmental risks could be placed into the "largely unfamiliar" category. Furthermore, if people believe that explosions could occur at factories, leading to the wide-scale spread of large amounts of toxic waste in the factory's surrounding environs, then people would probably recognize such risk to be "scary." Accordingly, just as is true with nuclear power plant accidents and genetic recombination technologies, environmental risk could potentially be perceived to be excessively higher than what is true in reality. Therefore, there is potential for local residents to not consent even if proper contamination policies are executed and the low danger of accidents are arbitrarily explained. Because of this, it is important for companies to carry out environmental risk communication with those who reside in a factory's surrounding environs.

Value of statistical life

In addition to examining countermeasures for environmental risk, it is additionally important to estimate the effects of reducing risk. For example, it is crucial to infer how the appropriate management of even a minuscule amount of toxic substances in drainage water could improve the risk of cancer related deaths among downstream residents. Furthermore, in order to compare the necessary costs of risk countermeasures, it is important to assess the effects of risk countermeasures in monetary terms. When the effects of measures to reduce the risk of death are assessed using monetary units, a guideline known as the *value of statistical life* is utilized. The value of statistical life is the quotient of the marginal willingness to pay for risk reduction divided by the scope of risk reduction.

Figure 6.3.2 explains the value of statistical life. The scope of risk reduction on the horizontal axis indicates the decree to which the risk of death is reduced through countermeasures. The vertical axis indicates monetary sums. The willingness to pay curve on the graph depicts an increasing willingness to pay to the extent that risk is reduced. Initial willingness to pay increases rapidly, but as people perceive the appropriateness of policies according to how much risk is reduced, the increment rate of the willingness to pay gradually decreases.

For example, let's assume that there is currently a death risk of 8/100,000. This death risk is to be decreased to 2/100,000 by enacting appropriate environmental policies. Upon this time, the scope of risk reduction could become 6/100,000, or (8/100,000 – 2/100,000). Here, assuming the willingness to pay for this environmental policy is 12,000 yen, the value of statistical life is equivalent to the 12,000 yen willingness to pay amount divided by the 6/100,000 scope of risk reduction, which amounts to 200,000,000 yen. This is the price per capita, so assuming there were, say, 7,500 people rescued from death through the execution of the aforementioned environmental policy, the result becomes the 200,000,000 value of statistical life × the 7,500 rescued people, amounting to 1,500,000,000,000 yen.

In order to estimate the value of statistical life, the scope of risk reduction and willingness to pay for a given environmental policy must be examined, and generally, the hedonic method (see Chapter 5 Section 2) and CVM (contingent valuation method, see Chapter 5 Section 3) are utilized. From here, consider a scenario where CVM is used.

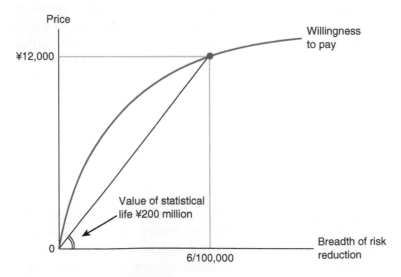

FIGURE 6.3.2 Value of statistical life

When using CVM, both the conditions pertaining to the risk of death currently, as well as the conditions pertaining to the risk of death after an environmental policy has been implemented, must be indicated to respondents. Then, the willingness to pay amount can be inquired regarding the change of the risk of death. However, even if the risk of death is simply indicated as "8 out of 100,000," it is not necessarily true that the respondent can appropriately perceive such risk, so it is important to devise methods for the respondent to easily understand the risk of death. In such instances, "risk ladder" and "dot" graphing methods are used to indicate the risk of death.

Figure 6.3.3 depicts a risk ladder used to indicate the risk of death. Based upon various causes of death risk, risk ladders place factors in a high position to the extent that

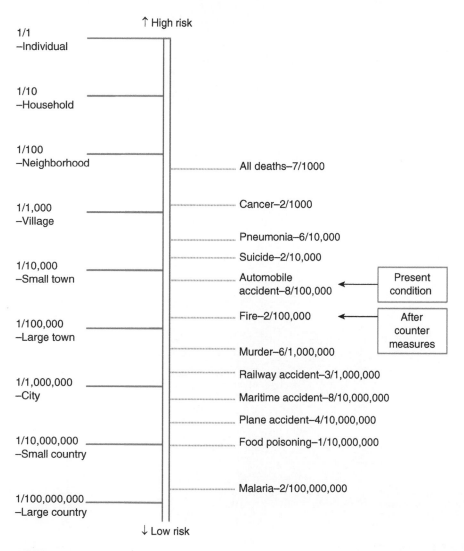

FIGURE 6.3.3 Risk depiction: risk radar

Source: Based on Statistics Bureau (Long-term Statistics of Japan) from the years 1990–1999

(A) 8 out of 100,000

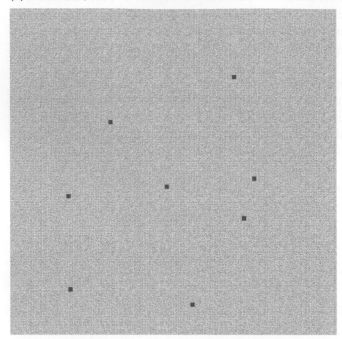

(B) 2 out of 100,000

FIGURE 6.3.4 Risk depiction: dot diagram

Source: Caution when making diagrams: Create a backdrop consisting of 100,000 boxes, and randomly fill in 8 for (A) and 2 for (B).

their risk of death is high, and conversely, they place factors in a lower position to the extent that their risk of death is low. By doing this, risk ladders succeed at juxtaposing various death risk factors, indicating their relative placement on the graph. For example, traffic accidents are accorded an "8/100,000" death risk, whilst fire accidents are equated to a "2/100,000" death risk.

Figure 6.3.4 depicts a dot method for indication death risk. On the left side, 8 out of the 100,000 squares on the grid are shaded in, indicating a death risk of "8/100,000." In the same way, on the right side, a death risk of "2/100,000" is portrayed. By comparing these two figures, one can visually confirm death risk reductions.

The willingness to pay amounts for reducing death risks are acquired using risk ladders and dot graphs to simply and accurately convey risk severity. Then, by dividing the willingness to pay amount by the scope of risk reduction, the value of statistical life can be acquired. There are numerous experimental studies assessing the value of statistical life taking place around the world. Based on an assessment value derived from a past research precedent performed by the U.S. EPA, the value of statistical life has been adopted as $4,800,000 as of the base year 1990, and it has been used to evaluate the *Clean Air Act* and other environmental regulatory policies. While there are only a few assessment precedents that have taken place in Japan, the Cabinet has placed a 4,600,000,000-yen value on traffic accident countermeasures, mirroring assessment values found overseas.

SUMMARY

In order to evaluate environmental risk based upon the probability of environmental contamination and the subsequent monetary damage amounts that would arise, businesses must have accident countermeasures in place before they occur. Since general people may perceive the risk of a contamination accident to be excessively large, businesses must create opportunities to perform risk communication that reflects the opinions and concerns of local residents while also communicating information about risk with the general public.

REVIEW PROBLEMS

1. Look up the environmental risks that a given business faces. Then, look up what types of risk countermeasures the firm has in place to cope with said risks.
2. Find an example in which risk perceptions vary between ordinary people and experts. Then, think about what the origins of such discrepancies are.
3. Assume that when the scope of risk reduction is 6/10,000, the value of statistical life is 200,000,000 yen, but when the scope of risk reduction is 2/10,000, the value of statistical life is 300,000,000 yen. Explain the discrepancies among these values of statistical life based on said risk reduction scopes.

LEARNING POINT: AN ECONOMIC EXPERIMENT WITH POLLUTION ACCIDENT RISK

Consider environmental risk with an economic experiment. Figure 6.3.5 lays out an economic experiment with pollution accident risk. First, create a six by six box grid, as displayed in the figure. Within 5 seconds, fill in circles in any squares on the chart of your choice. The number of circles does not matter. Assume factories are placed where the circles are, and each circle leads to profit earnings of ¥1 billion. Next, pollution is yielded at four locations. Roll a die twice, and fill in the corresponding boxes on the grid with red ink. For example, if the first roll results in a 5, and the second roll results in a 2, then fill in square 5–2. Repeat this four times to determine the four locations that pollution is yielded (if the same numbers appear, re-roll and select a separate location). If the pollution area has a circle in it, then the factory there is causing the pollution damages. Damage costs for each factory that produces harmful contamination amount to 10 billion yen. Under these conditions, how many factories ought factory owners to construct?

Now, if only one circle were placed on the grid, then the factory count would be one and profits would amount to $1 \times 10 = ¥1$ billion. Since pollution is yielded at four locations across the 36-square grid, so the probability of an accident occurring in the circle-marked box is 4/36. Since the ¥10 billion in damages arise when accidents take place where there are factories, the damage risk when there is only one factory is 10 billion yen $\times 4/36 = ¥1.11$ billion. That is a ¥110 million deficit. When there are n factories, the profits are $10 \times n$-hundred million yen. On the other hand, regarding pollution accidents, since the probability of an accident occurring in a factory zone is 4/36, when there are n factories, the likelihood is $n \times 4/36$ times. Accordingly, the damage risk from a pollution accident is 10 billion yen $\times n \times 4/36 = 11.1 \times n$-hundred million yen. By subtracting this, the deficit becomes $1.1 \times n$-hundred million yen. In other words, the more circles there are,

1. Create a 6 x 6 square grid.
2. Factory distribution: within 5 seconds, fill in circles into boxes of your choice. (One circle amounts to ¥1billion in profits.)
3. Contamination Accident: Incidents occur at 4 locations. Roll dice to determine where such accidents occur. Incidents occurring at places where factories are located result in ¥10 billion in damages.
4. Calculate final profits by subtracting damages from profits.

Profit: ¥5 billion (5 circles)
Damages: ¥10 billion (accident that took place at one factory location)
Difference: –¥5 billion

FIGURE 6.3.5 An economic experiment with contamination accident risk

the larger the deficit becomes. Additionally, if the number of circles were 0, there would be 0 factories and 0 profits accrued, and the damages caused by the firm would be the smallest possible. In short, with this experiment, the most appropriate decision for the business owner would be to not construct any factories at all.

Section 6.4: The ecosystem and biodiversity

What is biodiversity?

It is said that a diverse set of 30,000,000 species of wildlife exist on Earth, each of which comprise a distinctive part of unique ecosystems across the globe. These diverse species have come to exist through the progression of a 4,000,000,000-year long evolutionary history. We human beings are blessed by so-called "ecosystem services," or the natural benefits that arise from existence of biodiversity among species (see Table 6.4.1). For example, the ecosystem supplies the food, water, timber, energy, and other resources necessary for life. Within the ecosystem also exist climate regulation, flood mitigation, water and air quality purification, disaster prevention, and other helpful natural functions. Many medical supplies are produced from materials that originate from plantlife, and many of the species that are currently not being used have the potential to be used to produce additional or new medical supplies in the future. Because of this, within biodiversity exists the value of genetic resources.

While humans receive many blessings from wildlife diversity in the above ways, development spurred on by economic activity worsens the degradation of the natural environment and leads to the extinction of wildlife, inducing rapid losses in biodiversity on a global scale. According to the United Nations Millennium Ecosystem Assessment, the current rate of extinction surpasses the natural rate of extinction of the past by 100–1,000 times, and it is estimated that the degradation of ecological services will continue to worsen. Once a species goes extinct, there are no artificial measures available to bring it back into existence. Because of this, while preserving biodiversity has become a pressing issue, because the effects of the loss of biodiversity are yielded on a global scale, preserving biodiversity requires international cooperation.

With this in the background, the *Biodiversity Protocol* concerning the use and preservation of biodiversity was adopted at the 1992 Earth Summit. The Biodiversity Protocol called for biodiversity conservation, sustainable use of biodiversity services, and just and fair distribution of the benefits earned from genetic resources. However,

Table 6.4.1 Ecosystem services

Supply services	Food, raw materials, energy resource provisions
Regulatory services	Climate regulation, flood control, waste processing
Cultural services	Recreation, ecotourism, scientific discovery
Foundational services	Nutrient cycle, soil formation, water and atmospheric purification
Conservation services	Biological diversity protection, disaster protection

there existed severely contrasting views between developed and developing countries. Developed countries emphasized that since the benefits acquired through the development of medical supplies were dependent upon genetic resources that were extracted from natural resources in developing countries, developed countries should be required to share those benefits with the developing nations. In the face of this, developed countries purported that developing new medicines require large amounts of investment, so they could not accept freely a share of the benefits gained from genetic resources with developing countries.

After this, with nations having ratified the Biodiversity Protocol, the Conference of Parties (COP) is held about once every two years in order to continue pertinent discourse. At the 2010 Biodiversity Protocol COP 10 hosted in Nagoya city, Aichi prefecture, Japan, the "Nagoya Protocol" concerning the distribution of benefits from genetic resources and the "Aichi Target" which aims as preserving biological diversity were adopted. Due to the widespread television and news coverage of biological diversity issues that coincided with the COP being hosted in Japan, domestic concern for biodiversity issues elevated rabidly throughout the country. Furthermore, the United Nations has depicted the 2011 to 2020 year gap as the "10 Years of Biodiversity," over which international initiatives for biodiversity will continue to gain prominence.

The economics of ecosystem and biodiversity (TEEB)

Since there are large costs required to preserve biodiversity, economic perspectives are indispensable. The start of "The Economic of Ecosystem and Biodiversity" in 2007 has served as a launchpad for garnering attention concerning the relationships between biodiversity and economics. The special feature of TEEB is that it accentuates the important role of economic policies and businesses in the preservation of biological diversity.

According to a publicly announced TEEB interim report in May of 2005, if the world proceeds as usual without implementing any sort of new policies, farmland conversion, development expansion, climate change, and other factors would lead to a loss of 11% of the environment that existed in the year 2000 by the year 2050. It also predicted a loss of 60% of the world's coral reefs by the year 2030. Furthermore, it would cost an annual $22,000,000,000 to preserve all wildlife that must remain living for ecosystem order.

In this way, large financial sums are required to preserve biodiversity. In particular, it is difficult for developing countries with weak financial bases to secure the financial resources necessary to preserve biodiversity on their own. Thus, in addition to being at the core of revising existing government-based conservation policies, TEEP is required for our economic activity structures to consider biodiversity and ecosystem services and to make the transition to a sustainable society.

For this, it is essential to construct a new market that appropriately assesses the value of ecosystem services that are currently not perceived. Thus, by allowing us to understand the costs inflicted on society through biodiversity loss via experimental studies up until now that have provided monetary unit assessments of the value of ecosystem services, TEEB has indicated the social significance of preserving the ecosystem and biodiversity (see Table 6.4.2). What has been emphasized is that it is necessary to construct a system for paying consideration to ecosystem services, as well as a system that can equally distribute the benefits of ecosystem and biodiversity preservation.

In this way, TEEP indicates the critically important role that businesses and economic policies must undertake in order to preserve biodiversity. As a means of putting fundamental theories into practice, an informational report was published for administrative authorities in November of 2009. In addition to thorough reviews of the subsidy policies that had been in place up until then, the report also indicated the importance of creating incentives for ecosystem preservation through the implementation of an ecosystem service payment system. Moreover, in July of 2010, an informational report aimed at businesses was published. This time, the critical role that businesses play in preserving biodiversity and the ecosystem were indicated. For example, compensation for environmental losses through economic activity, placing financial burdens on businesses for new forms of environmental restoration, and other such biodiversity banking systems were introduced. TEEB draws attention to the new business opportunities of biodiversity preservation for these kinds of firms.

Payment for ecosystem services (PES) systems

Humankind and other life forms receive many benefits from biodiversity. For, forest ecosystems, provide multiple natural services, including landslide prevention, water source conservation, wildlife protection, and countermeasures to climate change through CO_2 sequestration. However, as no market prices exist for the majority of these ecosystem services, and there is also no system to impose a financial burden on those who benefit from said services, preserving the ecosystem is in no way tied to profits.

Table 6.4.2 Monetary value of ecosystem services

Item	($/ha/year)				
	Total	Supply service	Adjustment service	Habitat service	Cultural service
Open ocean	9	0	7	2	1
Coral reef	206,873	20,078	186,795	0	0
Coast	77,907	1,453	76,144	164	146
Coastal wetland	960	0	960	0	0
Inland wetland	282	167	115	0	0
Rivers and lakes	812	3	129	681	0
Tropical rainforest	29	0	12	17	0
Temperate forest	1,281	3	1,277	0	0
Forest land	5,066	25	130	1,005	3,907
Grassland	752	0	752	0	0

Therefore, "Payment for Ecosystem Services (PES)" systems in which those who receive benefits from ecosystem services pay for such services are being implemented worldwide. Across the globe, it is said that over 300 PES cases have been implemented, and PES is drawing further global attention. PES can exist in one of three forms (see the **Learning point** of this section):

1. Systems in which those who receive benefits from services voluntarily contribute funds,
2. Systems through which the government is the main source of funding, and
3. Systems that resemble PES in structure.

Biodiversity offsetting

The need for businesses to consider biodiversity as they perform economic activities is rising. As businesses develop, ideally, they should avoid influencing or reduce their effects on biodiversity to the greatest extent possible. However, is it true that there are also effects yielded that cannot be avoided? In these types of circumstances, we could imagine paying reparations for environmental losses by restoring nature in so me other location. For example, when wetlands are lost due to road development, we could imagine creating a new, similar wetland in a neighboring area. This kind of compensation mechanism is known as *biodiversity offsetting.*

Biodiversity offsetting had been implemented among environmental assessment systems in the U.S. decades ago. It could also be viewed as a compensation measure through which business entities voluntarily implement environmental restoration projects, or even as compensation measures through which financial burdens for environmental restoration are attributed. Because of this, even in instances where business operators cannot independently preserve biodiversity, the advantage arises from cooperating with other companies in biodiversity preservation aspirations. However, as one premise for implementing biodiversity banking, ecosystem services lost through development and the newly created ecosystem services for environmental restoration must be of equal value, and there must be no net loss in overall ecosystem service value.

In the U.S., methods for assessing the ecosystem known as "Habitat Evaluation Procedures (HEP)" are frequently used. These are processes through which wildlife habitat assessment targets are quantitatively measured from substance, space, and time perspectives. While HEP allow for relatively simple measurements to take place, it is difficult to evaluate ecosystems of differing natures. For example, when a forest is planted as a replacement for lost wetlands, it is difficult to determine the compensation measures when evaluating the different natural qualities of wetland and forest ecosystems using HEP. In order to make comparisons among different ecosystem services, such ecosystem services must be evaluated monetarily.

Furthermore, in recent years, international frameworks for biodiversity banking have gained attention. For example, the "Business and Biodiversity Offsets Programme (BBOP) sets the standard for the proliferation of biodiversity offset programs. In addition to a host of participating international organizations, national governments, and

NGOs, many multinationally active large resource development corporations are also involved in BBOP. Within this framework, developing countries are recognizing how essential it is to consider biodiversity when developing resources, and as such, the validity of biodiversity offsets is being recognized. While it is predicted that in the future, developed countries will come to pay even greater attention to biodiversity offsetting among developing countries, it is essential to develop methods for developing countries to appropriately evaluate their ecosystem services.

SUMMARY

Since the costs to preserve biodiversity are so great, international initiatives are indispensable. Various economic methods for preserving biodiversity, from "Ecosystem Service Payment Systems" in which those who benefit from ecosystem services pay for them, to "Biodiversity Offsets" through which ecosystem destruction that arises through development is compensated with the restoration of new natural environments, continue to be implemented.

REVIEW PROBLEMS

1. Look up the forest taxes implemented across the country, and consider the advantages and problematic points of each.
2. Investigate issues with implementing domestic biodiversity offsetting.

LEARNING POINT: PAYMENT FOR ECOSYSTEM SERVICES (PES) SYSTEM

1. Precedent of voluntary funding: Paying for ecosystem services provided by Vittel

   In the 1980s, Vittel brand natural mineral water was faced with the problem of deteriorating water quality at fountainheads due to a flourishing livestock industry near water sources in northeastern France. Cooperation with farmers was essential for water quality restoration, but if pesticide use were prohibited, then agricultural productivity would abate, leading to skyrocketing damage costs for farmers. Vittel and water source-area farmers entered negotiations, and Vittel agreed to supply funding for the farmers' water quality countermeasures. Vittel paid 24,250,000 Euros over a seven-year period for the water source countermeasures.

2. Precedent of government-based payments: Costa Rica's Forest Conservation System

The government of Costa Rica implemented a PES system in 1997 as a measure for preserving biodiversity. The Costa Rican PES system was a structure under which the government supplies funding to landowners for forest preservation. It involved three contract types:

  a. Forest conservation contracts (210 USD/ha),
  b. Sustainable forest management contracts (327 USD/ha), and
  c. Reforestation contracts (537 USD/ha).

Through these contracts, funding was allocated to the extent that forest preservation was successfully carried out, incentivizing voluntary forest preservation measures among landowners. However, the problem with this kind of PES is that financial reserves must be available for government-based payouts to be made.

3. Precedent of a structure similar to PES: Japan's Forest Environment Tax

The Forest Environment Tax is a system through which the forests' environmental preservation functions are safeguarded through forest maintenance activities that are funded by levying taxes on citizens who ultimately benefit from healthy forests. Taking the implementation of forest environmental tax in Kochi Prefecture in 2003, forest environmental taxes began to be implemented nationwide, and by 2012, 33 local governments had begun utilizing them. The financial burden amounts to 500–1,000 yen per person per year. The forest environmental tax pays for ecosystem services by levying a tax for the costs of forest management activities on the citizens that benefit from them. However, since there exists no structure for paying values to the extent that forest maintenance is performed, there is little incentive to preserve biodiversity. Furthermore, in a great number of cases, the basis of environmental service value is insufficient, and levied fees are not determined to the extent that benefits are received.

CHAPTER 7

Global environmental problems and international trade

Chapter overview

This chapter provides a number of perspectives about the long-term effects of global environmental problems and their countermeasures. One of these perspectives concerns the inexorable link between globalizing economies and the environment, as well as how the former impact the latter. Other perspectives consider the potential impacts of technological development and the principles of sustainable development. This chapter begins with an explanation of the impact that globalizing economies have on the environment and describes how implementing environmental policy in one country could have broader, international ramifications.

As technological development could potentially reduce environmental policy costs, this chapter outlines the technologies that could play major roles in achieving such environmental preservation measures. And, considering the recurring global debates regarding the impact that current generations will have on future generations, this chapter concludes with an explanation of how sustainable development is indispensable to responsibly addressing the world's most pressing environmental challenges.

Chapter content

Section 7.1—This section focuses on rapidly expanding global economies and whether globalization has a positive or negative impact on environmental conservation. Then, it explains how a given country's domestic environmental regulations affect both domestic and international economies.

Section 7.2—Advanced technology is a key component of coordinated efforts to tackle environmental problems. This section provides a number of voluntary environmental policy precedents among businesses and explains how environmental regulations could potentially spur on productivity improvements or even discourage advanced techno-logical proliferation.

Section 7.3—The concept of sustainable development, often cited as a key tenet among international council reports concerning the environment and develop-ment, is central to debates on how to achieve economic growth and environmental conservation simultaneously. This section provides a definition of sustainable devel-opment and examples of environmental Kuznets curves for potential sustainable development.

Section 7.4—This section concerns the supply and demand of world energy and the history of renewable energy and electricity and gas industry deregulation in Japan.

Section 7.1: International trade and the environment

Globalization and the environment

There are many conflicting opinions about rapidly globalizing economies. On the one hand, there is an optimistic view, which points to the expansion of competitive markets that have transcended national barriers, leading to increased trade and foreign direct investment. According to this view, such activities have allowed for greater transcon-tinental flows of technology, increased employment opportunities, and improvements in the overall welfare of human beings alongside economic development. However, on the other hand, there also exists a pessimistic view, which views economic globali-zation as a catalyst for environmental destruction and poverty.

The 1991 yellow fin tuna incident between the U.S. and Mexico is a prime example of an issue involving both trade and the environment. At the time, U.S. *Marine Mammal Protection Law* ensured the wellbeing of dolphins. Adhering to this, the U.S. prohibited Mexican tuna imports based on the fact that Mexican tuna was produced using fishing techniques that yield catch loads with high percentages of dolphins mixed in (i.e. the amounts of dolphin and other non-tuna fish among catch loads were quite high). Shortly thereafter, Mexico brought forth a claim to the General Agreement on Tariffs and Trade (GATT, an international pact aiming to promote market liberalizations across the world), submitting a panel report alleging that the actions of the U.S. went against the GATT objectives. In the face of this, the EPA expressed strong dissent, and the issue developed from a trade and environ-mental issue to political strife.

Later, the *Rio Declaration on Environment and Development*, and *Agenda 21* were adopted at the 1992 Earth Summit. It was at this point that the international community really began to recognize the importance of simultaneously implementing trade and environmental policies. In this unit, we consider various viewpoints about the relationship between globalization and the environment. Particularly referring to the relationship between the environment and trade, we expound upon what type of influence environmental policies of one country have on trade within the context of the conflicting aims of market liberalization and environmental preservation.

The impacts of trade on the environment

Up until now, we have seen how information is disseminated through trade, making trade an economic development-inducing factor. However, as economic activity becomes more vigorous through trade liberalization, debates surrounding the exact nature of the impacts of trade on the environment gain more weight when drafting trade policy. While trade liberalization could harm the ecosystem through greater economic activity, it could also function to restore the environment as a catalyst of technological change. In other words, be it negative or positive, trade could not be said to have singular or insular impacts. It follows that considerations of trade's impact on nature should not only include aforementioned direct effects, but also such indirect effects like upswings in environmental awareness as wealth accumulates throughout the general populace. In general, the effect of trade on the environment can be divided into three categories: *scalar effect*, *technological effect*, and *structural effect*. It is crucial to include these three effects in policy debates.

The first of these, which increases environmental pollution alongside economic activity, is known as the *scalar effect*. It is the effect by which the increases in production through trade liberalization lead to increased pollution levels, and it is considered to underlie negative environmental impacts. Moreover, this effect is a key component of the argument against trade liberalization, which exacerbates environmental destruction. Be that as it may, could one really say that trade and the environment are fundamentally at odds with each other?

Next, consider the impacts of increased income through trade. Said scenario paves the way for an introduction into the second effect of trade, known as the *technological effect*. This includes the manner by which changes in production processes affect pollution. More specifically, the technological effect is one by which pollution emissions content per single unit of production decrease through improvements in production processes (i.e. technologies). The result of this is that as incomes increase alongside economic development, environmentally friendly materials and equipment become more commonplace. And, in addition to these technological innovations which yield less environmentally harmful products, as income levels rise, the concern for environmental wellbeing held by citizens increases. This serves as a driving force for communities to reassess environmental laws and regulations. For example, demand for environmental goods increases as environmental regulations are implemented, all of which alleviates damage to natural environs. It is no surprise, then, that the technological effect considered to be at the crux of the wage increases that have favorable impacts on environmental conditions.

The third and final effect is known as the *structural effect*. This, with processes for economic development, represents changes in the domestic component distribution ratio of goods production that have negative impacts on the environment (i.e. pollutants) and those goods that do not (non-pollutant goods). For example, we know that environmental damages inflicted by farming and textile industries have increased as they have shifted from a labor-intensive stage where not many pollutant resources were used to a more energy- and capital-intensive stage where pollutants became more common as their industrial structures changed. Furthermore, service and IT industries have displayed a reverse trend since their formation, decreasing their burden over time. When temporary, non-conventional changes in operations and other aspects of primary industries (farming, fishing, etc.), secondary industries (manufacturing, construction, etc.) and tertiary industries (service industries, etc.) change, so too does the proportion of domestic components that are distributed among produced goods, in turn elucidating the presence of both positive and negative impacts on nature. As such, the structural effect is considered to be both beneficial and harmful to the environment.

The impact of comparative advantage

The next task is to assess trade's environmental impact with regards to industrial structure. Countries that have a *comparative advantage* in environmentally harmful goods (e.g. China, which produces a vast amount of atmosphere-polluting coal, or any other country that has an abundance of such factor inputs) deal damage to natural cycles when they specialize in such goods as markets liberalize. And, while the majority of those dealing environmental damage are developing countries, people still claim that their industrial structures are at odds with moving towards market liberalization. How could this be so?

Imagine a country making a transition from capital-intensive industries with comparatively low technological standards to high-level, technology-intensive industries that require abundant labor resources. Trends show that when there are many highly natural resource-dependent industries in developing countries, greater numbers of comparatively clean, environmentally friendly industries are established in developed countries. In turn, pollution-intensive industries pop up throughout a host of developing countries, giving rise to unavoidable environmental destruction.

Yet, is it true that industries with the comparatively worse pollution trends exist predominately in developing countries? Take for example steel, nonferrous metal, oil purification, and other such capital-intensive industries that are relatively energy- and natural resource-dependent and which are also pollution-intensive and emit harmful substances in vast quantities. Many of these industries are based in developed countries that have generally acquired comparative advantages of significant capital. Contrastingly, labor-intensive industries pollute very little as they consume very few resources in their processes. Since developing countries are centered on labor-intensive industries, trade propagation poses the risk of inviting or supporting the growth of environmentally harmful industries as well. This is known as the *factor endowment effect*. In this case, developed countries that host a great number of pollution-intensive industries experience the symbiotic nature of exports and pollution, inevitably resulting in greater production levels. On the other hand, developing countries with many labor-intensive industries export greater amounts of comparatively environmentally friendly goods, resulting in environmental restoration at the domestic level.

However, an important consideration regarding the factor endowment effect is that developed countries adopt more strict environmental regulations. Known as the *environmental regulation effect,* this is one outcome of greater environmentally conscious manufacturing processes among these countries.

As seen above, as environmental preservation and trade liberalization are fundamentally at odds with each other, achieving both is no simple feat. Nevertheless, considering the importance of their co-implementation, knowledge of construction, and operational processes among various firms is of paramount importance.

The effects that environmental policy has on trade

Environmental dumping

Shifting focus from the effects that trade has on the environment, this section serves to demonstrate how a particular country's environmental policies affect trade. Specifically, the effects that strict environmental policy has on a country's domestic businesses, as well as the effects that the same policy has on foreign enterprises, are discussed.

The first pertinent concept to be discussed is an issue known as *environmental dumping*. When environmental regulatory standards differ between countries, eventually, firms that manufacture goods in countries with relatively loose environmental regulations, compared to firms that operate in countries with strict regulations, can produce without such stringent production cost burdens of adhering to regulations. Therefore, the former will enjoy a competitive edge in international markets since they are burdened less by environmental policy costs. In this scenario, in addition to diminished competitive power among firms operating out of countries with strict regulations, firms operating in countries with loose environmental regulations are able to export their goods at cheaper prices. *Dumping* is the practice of vastly decreasing export prices and driving away competitors in export destination countries. This phenomenon, when related to the environment, is known as *environmental dumping*. The North American Free Trade Agreement (NAFTA) has debated the environmental dumping practice in judgments regarding signatory countries. Mexico has been cited for its capacity to produce comparatively cheaply and export to the U.S. due largely to its looser environmental regulations. Elsewhere in the world, northern European environmental taxes and the U.S. renewal of the *Clean Air Act* at the start of the 1990s would impose huge financial burdens on industries with international comparative advantages, leading governments to address the related dumping issues.

One pertinent historical trend relates to U.S. companies that, while complying with NAFTA regulations, choose to relocate their factories to Mexico and take advantage of cheaper production where environmental regulations are less stringent. From there they are free to export their goods to the U.S. In a similar fashion, many Japanese companies shifted their factory operations to China and other Southeast Asian countries when domestic environmental regulations grew tighter.

The loss of employment that arises in developed countries that had originally hosted manufacturing plants after companies shift their factory operations to developing countries and export their former domestically made goods is not the only issue at hand. Another issue is that firms that emit large pollution quantities can simply transfer to countries with looser environmental regulations in cases where host countries implement regulations that are too taxing. From the standpoint of the firm that owns the polluting factory, shifting operations to a country with comparatively looser environmental policies in order to avoid strict regulations is known as the *pollution escape effect*. Moreover, globalizing economies allow companies to shift their central production locations relatively freely, and many developing countries aim to draw in firms that could contribute to their economic growth through industrial activity. To this extent, many fear that such developing countries would actually compete amongst each other by loosening their environmental regulations in order to attract more businesses.

Additionally, many developed nations maintain strict disposal-related industrial waste regulations. As a result, even if countries do not shift their operational base location, they nevertheless remain unchecked from exporting toxic waste substances. Thus, policies that were originally ratified to protect the environment have effectively caused a shift in the burdens of hosting toxic substances from the countries in which they originate to others (particularly to developing countries). This suggests that the one country's environmental regulations could be completely unrelated to the overall global emissions reductions.

Barriers to trade due to environmental regulations

It is essential to firmly grasp the extent to which one developed country's exports affect foreign firms in cases where the former tightens its environmental regulations. Prime examples of this include legal prohibitions on the sale of non-renewable containers, as well as import restrictions on products made with detergents that damage the ozone layer. Many of these cases adhere to the environmental observation and management rules established by the International Standardization Organization (ISO). When newly established environmental rules are systemically integrated into the firms in a given country, they are inherent aspects of all transactions between the domestic firms and their overseas counterparts. Firms that do not comply with the rules are banned from business, while those that do inherit enormous costs that act as barriers to trade.

So, is it safe to say that the environmental regulations of one country necessarily have negative impacts on trade with another country? This concept is laid out in in more detail in the final section. For now, note that during the research and development stage of strict environmental regulation, as new technologies diffuse transnationally, productivity improves in the long run. Thus, just as free trade can either bolster or undermine environmental conservation, environmental regulations can similarly have positive or negative impacts on trade. What's more, with globalizing economies in the backdrop of recent affairs, one nation's environmental policies have come to have unexpected side effects, particularly with regards to the nature of impact of one country's trade and investment schemes on other countries. Therefore, when a given country adopts environmental policies, it must also consider the greater effects that they will have internationally.

Effects of trade restrictions

Next is an explanation how enforcing trade restrictions in the context of international environmental protocols (see the **Learning point**) affect markets and the environment differently. The Washington Convention is a prime example of these trends. Formally known as the *Washington Convention on International Trade in Endangered Species*,

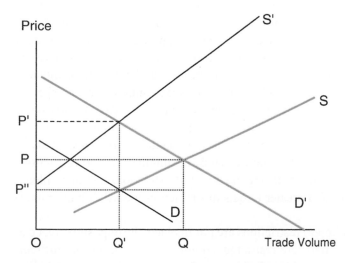

FIGURE 7.1.1 Impact of import restrictions on transaction prices

this convention aims to protect wildlife on the verge of extinction by restricting cater-ing and procurement by incentivizing cooperation between exporting and importing countries via international trade regulations for wildlife species.

Moreover, the convention set international trade regulations according to the extent to which wildlife species face the threat of extinction. Such regulations include a permit system along which each country must monitor its own wildlife imports, and they are divided into three categories. The first is the default prohibition of commer-cial trading of endangered wildlife. In order for such dealings to take place, partici-pating countries must acquire import and export permits (depending upon their role in the transaction). The second class involves wildlife that are not currently endan-gered but could still go extinct in the absence of stringent regulations. In cases such as these, commercial dealings are possible if exporting countries acquire appropriate licensing. The third and final category involves species that are not in immediate danger, but still need to be accommodated under the transparency of international cooperation.

Note that the Washington Convention precepts are in line with World Trade Organization (WTO) rules. Wildlife, like scarce resources, is dealt with out-side of the realm of free trade under trade restrictions. Consider the supply and demand of a given wildlife species as depicted in Figure 7.1.1. The supply curve represents the global supply of the species. And, while the demand curve depicts total global export demand, yearly transaction price P and quantity Q are located at the intersection of the two curves. When trade restrictions are enacted over a specified one year time frame, the quantity traded shifts to Q'. From here, we compare the effects that export and import restrictions of the same quantity have on price.

When exports are restricted, they are more difficult to carry out, so the price for the same transaction quantity escalates. Figure 7.1.1 displays a supply curve shift from S to S', and a price increase to P'. In most cases, property rights for endangered spe-cies are not established, so many wildlife catching and trapping techniques, both legal and illegal, take place. What can be understood from this is that when prices increase, people are incentivized to catch and collect greater numbers of the then more valuable species. That is, higher prices spur people to acquire greater numbers of the species for greater gains. Further still, since there are limited numbers of said organisms, com-petitors take on a heightened sense of risk as they endeavor to preemptively catch as much of the wildlife as they can in order not to miss out on the narrow opportunity to earn greater profits through sale of the wildlife. Thus, in summary, export restrictions raise the transaction prices of wildlife facing the threat of extinction, leading to illegal or black market transactions and greater threats of extinction for wildlife in unman-aged markets.

On the other hand, when import restrictions are put in place, the opposite effect can be witnessed. Namely, as imports are restricted, equally priced imports are difficult to obtain and the demand curve shifts downward. Trade quantities set at Q' align with prices that have dropped to P'. Unlike the export restriction scenario, the price has shifted in the opposite manner, leading to an opposite effect. In other words, since the marginal benefits (e.g. profits from sale) of capturing or collecting wildlife diminish, the risk of over-capturing also abates. Thus, although the goal of these trade restrictions is environmental protection, since wildlife species procurement is unrestricted, global transaction prices for importing of species drop, and preservation is effectively achieved.

SUMMARY

International environmental problems arise alongside globalization. As markets continue to liberalize, various positive and negative impacts on environmental preservation come into existence. In many cases, a country's environmental regulations exert influence on economies, both domestic and abroad. It is critical that the roles of international protocols aimed at promoting market liberalization are made compatible with those of international environmental agreements.

REVIEW PROBLEMS

1. Explain the impact that international trade yields on the environment.
2. Describe the impact that one country's environmental policies could have on another country's trade.
3. Summarize the trade restrictions that are pertinent to environmental protection.

LEARNING POINT: INTERNATIONAL AGREEMENTS FOR TRADE AND THE ENVIRONMENT

There is an abundance of dialogue regarding globalizing economic activities, particularly those related to the environment. At the GATT Uruguay Round hosted from 1986 until 1994, while the focus was not on direct negotiations for environmental problems, environmental considerations became internationally standardized, and "environmental safeguarding and conservation, as well as sustainable development" were announced as the pretexts for establishing the World Trade Organization (WTO) as the GATT's successor organization. Furthermore, in order to further investigate these problems, the "committee for trade and environment" was instituted at the Marrakech Cabinet meeting. However, establishing this committee, as well as the Earth Summit agendas, render resolutions to trade and environmental issues simple. For starters, the goals of the GATT/WTO were to promote international trade liberalization by removing trade restrictions from all countries.

Up until then, there existed numerous multilateral environmental agreements for environmental safeguarding and conservation across multiple countries and even at the global level. And, with the MEA, there were cases where trade restrictions were imposed on signatory countries responsible for environmental problems. The Washington Protocol, the Basil Protocol, and the Cartagena Protocol are all prime examples. Conversely, the modern WTO has no stipulations for direct regulations related to the environment. In other words, it is essential to realize that the WTO, principally aiming to maintain and expand free markets, does not have the same aims as the MEA.

Consider the Convention on Biodiversity (the Cartagena Protocol for Biosafety) based on the Cartagena Protocol as a precedent. Here, import and usage restrictions on genetically modified organisms were implemented to protect against said organisms' impacts on health and the ecosystem.

Contrary to import restrictions imposed depending on the safety checks of products based on the Cartagena Protocol and performed by importing countries, with WTO rules, importing countries endeavor to acquire scientifically based safety guarantees. Where that is not possible, imports are prohibited.

While there may exist conflict with consistency between the two agreements, when there is no discrimination among unilateral trade restrictions in place, they are viewed to adhere with WTO rules. This is because in such complex cases, international agreements related to trade have historically been placed ahead of environmental issues in their level of importance. Therefore, in the future, in order for the benefits that countries receive from participating in both international trade and new *international environmental agreements* to not be at odds with each other, initiatives for solving international disputes must be strengthened among all parties involved.

Section 7.2: Environmental regulations and technological advancements

Economic development and productivity

In the wake of the post-WWII economic growth among developed countries and the more modern growth among developing countries, environmental problems and social losses have exacerbated, prompting greater adoptions of environmental regulations. In general, environmental regulations are believed to lag behind economic development. This section discusses how economic development and productivity relate to environmental regulations. Economic development is determined by the degree to which labor increases, capital is acquired, and whether or not productivity increases. Capital is the sum of assets owned by a given company, and the degree to which a company enhances its productivity and labor largely influences the speed at which it acquires capital. For instance, a company that can increase its productivity can also increase its revenues and thus proceed with acquiring greater capital. And, said company would aim to do so while also restricting rent per capita as the number of laborers increases and increasing investments in its facilities, thus further accumulating capital.

Labor unions, relatively unable to put income to use effectively, are mainly concerned with changes in productivity. *Productivity* is the ratio of investment to production (amount of production/amount of investment). To the degree that output is greater than investment inputs, productivity is enhanced. Such production components as labor, capital, land, raw materials, fuel sources, and mechanical equipment all represent investment levels, while production amounts, profits, and GDP are all production levels. When productivity rises, comparatively cheaper production, additional labor, and greater profits all become possible. Furthermore, production efficiency underpins a company's profitability, and productivity is a measure of the quality of a firm's production efficiency. Productivity is an across-the-board indicator of innovation and technological improvement.

In the 1960s, Japan improved its productivity so much that it was among the best of developed countries. However, from the 1970s, investigating the means to further enhance production speeds made its way to the top of the agenda throughout Japanese firms. Environmental regulation fortifications of the time were thought to function as catalysts of these aims. Few though there were, in the short term, environmental regulations increased capital (pollution purification apparatuses) and labor (pollution purification operations) that were not tied to the company profits became sources of increased costs. Even the smallest increases in investments in the denominator, depending on the scenario, come with price increases that result in declines (e.g. production levels) in numerator profits, which would lead one to believe that there is a negative impact on productivity. In other words, attaining greater productivity and environmental protection were thought of as inconsistent with each other. The aim of this section is to confirm whether this is true over both the short and long term.

What is productivity?

From here, it is important to gain an understanding of productivity and its influence on international competitive power. When environmental regulations are strengthened at the single-country level, they could signify that country's weakened industrial competitive power at the international level. International industrial competitive power is a comparative strength that is depicted in free international markets. It is a country's production capability with relation to its ability to increase export shares. However, since exports fluctuate largely due to exchange rate adjustments, international competitive power is determined, at least to some degree, by outside factors. Thus, it is important to pay attention to the fact that international competitive power is not solely related to the rise of productivity, which is the ratio of investment levels to productivity levels (amount of production/amount of investment). In fact, the components of international competitive power are so ambiguous that this section would be hard-pressed to explain them all. Rather, priority will be placed on analyzing how the economic problems that countries face impact productivity improvements. And since not only export sector, but also domestic sector, production influences the national standards of living, one must pay close attention to whether or not domestic economies are efficient and whether or not productivity is improving.

Figure 7.2.1 takes 1990 as a standard year for comparison and indicates investment (e.g. capital, labor) and production levels. It makes comparisons with the year 2000, where investment levels increased by 10% (from 100 to 110) and production levels rose by 20% (from 100 to 120). In this scenario, the difference in rates increases is 10% (= *20% − 10%*), which indicates increasing productivity.

Moreover, when thinking about productivity in this manner, not only entire-country analyses but intra-company assessments of technological improvement, where possible, are useful. The European Council hosts debates about post-Kyoto Protocol aspects with regards to environmental problems. Technological advancement is now a central theme of the talks, primarily because solving environmental issues through such methods as reducing CO_2 emissions relies heavily on technological improvements. Yet of equal significance is the fact that improvements in technology tend to bolster productivity.

Technological proliferation

Components that obstruct technological proliferation

While the technology diffusion greatly impacts productivity, there remain many obstructive factors to *technological proliferation*. Consider a scenario where it is possible to reduce fossil fuel expenditures due to the presence of highly energy-efficient systems. In this case, energy produced could feasibly cover investment costs in the

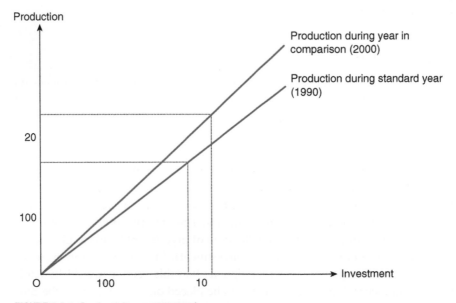

FIGURE 7.2.1 Productivity measurement

short term. However, over the long term, technological proliferation slows as the phenomenon known as the energy paradox comes into play. Why, exactly, does it occur?

The first point of consideration used to shed light on this question is the immense amount of uncertain effects that arise as new technology is used. For example, it is safe to assume that when new technologies are implemented, investment costs could be returned over the medium term. However, the rate of these investment returns is inherently unknown. In such scenarios, waiting for another business to implement new technologies is often a preferable approach, as one company could use others' successes and failures to guide its technological advancement.

The second point concerns the benefits from energy conservation that firms gain as they utilize newer technologies. Those who can afford to invest in new facilities and equipment anticipate many advantages to possessing the most up-to-date technologies. However, problems arise when those who cannot afford to make such investments must rely on those who can. That is, the relationship between representatives (agents) and dependents (principles) with unique interests is inherently problematic in that the former do not necessarily act in the most ideal manner from the viewpoint of the latter. In fact, it is fairly uncommon for the interests of both parties to be in line with each other. This is known as the principle/agent problem. Consider, for example, a case where residents (principles) of an apartment complex request that the building owner (agent) install air conditioning in their housing units. As the cost for the owner to install a high-efficiency energy system would be so great that residents would be unable to pay rent, the owner is left with no incentive to grant the residents' request.

The third point is that even when cutting-edge technologies are installed, to the extent that they are advanced, greater numbers of laborers who possess specialized knowledge of the new technologies become indispensable (which could raise labor costs). Therefore, if a firm does not already command technology of a certain level, making the jump using more sophisticated technologies could cause implementation to be hindered or even abandoned.

Components conducive to technological proliferation

Since problems arise at various stages of a new technology's diffusion, the degree to which technology is upgraded for the aims of effective environmental protection is a significant criterion for judging the success or failure of environmental regulations. Technological advancement is of supreme importance because short-term innovations that lead to pollution reductions without cutting cost increases could also produce savings on component costs that alleviate financial losses in the long term.

Strict environmental regulations are thought to negatively impact productivity because even in the short-term they become a source of rising costs for firms. Since the social costs of environmental pollution are reflected in the costs of creating goods and services, environmental taxes that are less stringent than direct regulations and other economic methods are considered to negatively impact productivity. Thus, to the extent that environmental regulations are strict, the monetary cost for mandatory, large-scale pollution reductions are also substantial. Driven by changes to labor and capital costs, firms seek to save on component costs by developing environmentally sensitive technologies through increased R&D investment. In other words, incentives for further pollution-curbing technology innovation arise over the mid to long term.

Moreover, when economic measures are applied via climate change policies, there are incentives to develop and adopt new energy conservation technologies. For example, the U.S., aiming to control acid rain, implemented an SO_2 emissions trade system. Participation was compulsory for electric power plants in every state outside of Alaska and Hawaii, and by the year 2010, SO_2 levels shrank to 50% less than 1980 levels. Thus, the system clearly led to improvements in de-sulfurization equipment efficiency.

Furthermore, by imposing carbon taxes, energy prices went up leading across-the-board replacements of older air conditioner, gas water heater, and central air conditioning units with newer ones. Indeed environmental taxes and other economic methods have spurred on some of the most noteworthy technological advancements. They have, for example, led to the development of ESCOs, de-sulfurization equipment, and solar energy. ESCO is an abbreviation for "energy service company," which provides a comprehensive service for energy conservation at factories and buildings, implementing environmentally friendly energy conservation services with proven statistics of energy efficiency.

Environmental regulations among businesses

End of pipe and *cleaner production* are two technology-based strategies that businesses employ. In recent times, companies have generally used end of pipe as a means to properly process pollution substances at points of emission, thereby reducing overall pollution. Through this strategy, production and material flow facilities are left unchanged, while emission points are equipped to more adequately process harmful substances. Depending upon the magnitude to which end of pipe is relied upon, it could contribute to greater operational costs and thus have a negative impact on productivity. However, on the topic of waste-processing, many other costs, including those incurred through recycling, are often greater than installing end of pipe equipment. Generally speaking, regulations inflate production-induced pollution costs considerably, in turn diminishing benefits. Yet, greater profits can also be earned when pollution prevention technologies are included in large-scale chemical and engineering plant construction plans. Looking at the economy as a whole, negative impacts have dwindled hand in hand with greater end of pipe technologies. Moreover, as many of these technologies were adopted as a result of widespread investments in pollution prevention in the 1970s, they are generally not considered to have negatively impacted Japan's economic development at the time.

Contrary to this, cleaner production encompasses both technologies related to specific policies as well as those pertinent to system management methods, spanning raw material acquisition to disposal and reuse stages. It includes all processes for reducing burdens on the environment. Cleaner production investment causes greater pollution reductions than end of pipe investments, which carries significant weight in an era where business process structures are being reconfigured with the aim of decreasing pollution. Furthermore, restructured production and business processes underlie production efficiency improvements.

The potential for environmental regulations to solicit productivity improvements

Production and business practices can be restructured based on environmental regulations, leading production technology levels to all-time highs. Thus, regulations inspire financially burdened companies to be innovative in their quest to cut pollution, which inevitably translates into productivity improvements.

For example, making existing regulations more stringent and implementing new regulations increases the incentive for owners of older factories to quickly upgrade facility assets. By doing so, the overall environmental damage abates, and moreover, firms gain renewed incentives to adopt state-of-the-art technologies in order to increase productivity. Just like in the previously mentioned energy paradox, pollution could potentially decrease as productivity rises. However, profits do not necessarily increase as regulations are adopted, and there are not always methods to avoid or soften monetary losses. Moreover, where emissions regulations for conventional and cutting-edge equipment throughout factories and offices differ, there is generally more incentive to utilize older equipment with looser regulations over newer equipment with stricter ones. This renders measurements of both productivity improvements and pollution reductions difficult.

Similarly, beyond not fully understanding the effects of continuing to use dated facilities and other information pertinent to conducting efficient business, additional problems arise from failing to capitalize on cost reduction opportunities (i.e. asset equipment upgrades, etc.). Take, for example, the results from a questionnaire survey directed at Japanese companies. According to the *Questionnaire Survey Report Regarding Environmentally Friendly Business Activities* published by the Ministry of Environment in 1999, over 60% of the companies listed on the stock exchange were unaware of environmental protection expenditures. The prevalent justification for this lack of awareness was that companies did not feel as if calculating such expenditures was necessary.

Regardless, 40% of the companies reported decreased environmental damages and greater-than-expected cost reductions through resource and energy saving measures achieved through attaining ISO 14001 certification (the international standard for environmental management systems). It is clear that addressing the inefficient circumstances underlying environmental problems is positively correlated with cost reductions. It thereby follows that resultant price reductions are as considerable as inefficiencies.

SUMMARY

One major criterion for assessing environmental regulations is the degree to which environmentally-friendly technology upgrades effectively reduce fuel costs through energy conservation. Of course, there are significant obstacles to promoting these technologies in the short to medium-term. Furthermore, it is possible for productivity to improve via environmental regulations when firms are highly inefficient.

REVIEW PROBLEMS

1. Explain why automobile pollution regulations spurred on technological improvements.
2. Describe factors that obstruct technological proliferation.
3. Describe factors that promote technological proliferation.

LEARNING POINT: AUTOMOBILE POLLUTION REGULATION

Automobile pollution regulations of the 1970s are often cited as precedents in which environmental regulations had positive influences on technological development. *The Musky Law* in the U.S., which mandated a 90% drop in automobile nitrogen oxide emissions, stands as a forerunner among similarly strict regulations. However, due to staunch objections from automobile manufacturers, the law was tabled until the mid-1980s.

Shortly following U.S. attempts, a Japanese version of The Musky Law (*Automobile Exhaust Gas Regulation Law*) was implemented in 1978, making it, numerically speaking, the world's most strict emissions reduction legislation. What followed was the birth of the CVCC engine, developed by Honda with the lowest-polluting engine technology of the time in order to clear regulatory standards. Other manufacturers would follow suit in their attempts to adhere to legal codes through the development of three-way catalyst devices, which were capable of purifying automobile emissions through oxidization and deoxidization. By the start of the 1980s, while U.S. automakers still struggled to implement exhaust gas countermeasures, their Japanese counterparts had already cleared those hurdles and had shifted their investment focus to improving gas mileage technologies for smaller cars. Naturally, these products swept through North American markets.

However, generally strict regulations do not necessarily yield technological development and productivity improvements. This is because there are often many other factors beyond the regulations themselves at play.

The first of these factors are the pertinent technologies that firms possess in reserve, developed even before regulations are enacted. For example, in addition to Honda (which, as previously mentioned, had successfully developed the first industry-leading CVCC engine), Toyota also positioned itself to pour large capital and personnel investments into developing a three-way catalyst in place of drafting new car models.

Yet, beyond specific firm endeavors to invent such competitive devices as those mentioned above, another factor that contributed to improvements in technology and productivity was the fact that the environment was steadily gaining weight as a salient political issue. Correspondingly, pollution problems and tightening emissions regulations grew to be more commonplace among political agendas. Within this context, the CVCC was hailed by the media as a step in the right direction towards resolving pollution issues. And, at the same time, a string of maverick mayors across seven major cities, each of whom emphasized their strong anti-pollution sentiments, willfully shared investigative reports on their capability to implement stricter gas emissions regulations, thereby contributing to the ongoing pollution reduction discourse of the day. In other words, not only the political regulations, but also the substantial demand for environmental friendliness, had major impacts on implementing strict regulations and applying constant pressure on firms to upgrade their technologies. Finally, encouraging business participation as a means to make the regulatory process clear and transparent from the outset of policy setting is also said to have contributed to the overall successes of pollution reduction policies.

Overall, stringent regulations in Japan that appealed to local government, consumer, and mass media demands for environmentally friendly policies gave rise to revolutionary technological advancements. Similarly, where other countries have been policy forerunners, they too succeeded by encouraging business participation at the initial stages of policy setting and including a system for adjusting policy terms over time. Thus, at the end of the day, overgeneralized, overly strict policies fail to address the complex pollution issues and instead lead only to greater production levels.

Section 7.3: Sustainable development

Principles of sustainable development

As they have been pushed into the limelight in recent years, there is an ever-growing exigency to tackle climate change and other environmental issues through global initiatives. Among these initiatives, assisting developing countries with their emission reduction endeavors is of prime importance. However, due to the potential side effects of stunted economic growth, developing countries are often reluctant to curtail climate change-inducing emissions.

Addressing this debacle naturally requires an understanding of sustainable development. Sustainable development entails using the environment to fulfill our current needs while also not damaging it in such a way that future generations would be unable to fulfill their needs through it. Currently, sustainable development is being recognized on the global scale as a fundamental principle of environmental preservation. In addition to promoting harmony between environmental cleanliness and economic development, it serves as a model for achieving development in moderation while putting forth reasonable efforts to protect natural systems.

This principle was first published in the *Global Environment Preservation Strategy* compiled in 1980 by such organizations as United Nations Environment Programme (UNEP) and the International Union for the Conservation of Nature and Natural Resources (IUCN). Subsequently, at the Earth Summit of 1992, the focus shifted to the *Rio Declaration on Environment and Development* and *Agenda 21* protocols, which, to this day, outline the fundamental concepts of multiple global environmental initiatives. The concept of sustainable society is, for instance, the premise underlying the 4th article of Japan's *Basic Environment Law* enacted in 1993.

Moreover, such ideals are central to the 1984 final report by the World Commission on Environment and Development (Brundtland Commission) titled *Our Common Future*, and become more broadly accepted thenceforth. In reports by the Brundtland Commission, the concepts were defined as "Development that allows us to meet the needs of the current generation without taking away from the ability of future generations to satisfy their needs."

The definition of sustainable development

Pareto fundamentals are said to be socially desirable conditions within which everyone's satisfaction levels increase without giving rise to disadvantages to anyone. The above definition of sustainable development provided by the Brundtland Commission alleges to perpetuate Pareto fundamentals across generations. It is difficult to depict scenarios involving problems with intergenerational equilibrium when the issue at hand is defining the most ideal utility distributions for both present and future generations. In other words, it is difficult to determine what should be passed on from generation to generation. Defining needs is also quite the challenge, as they differ with each person or group of people. With all of these aspects in mind, let us reconsider what the most appropriate definition of sustainable development could be.

From the outset, the concept of "development" ought to include real per capita income, health and nutrition conditions, education, access to resources, equality of income distribution, fundamental freedoms, and other specific aspects. And, sustainable development is assumed to pass all of these indicators across all periods without causing any of them to diminish. Yet, in cases where some of the aforementioned, specific indicators are referred to, justifying why other indicators are not of equal importance is a challenge. And of course, the greater the number of indicators, the more problematic it becomes to determine an accurate, efficient definition. Thus, there are many hurdles to outlining sustainable development conditions and indicators. Finally, in cases when one or more indicators worsen upon the improvement of another or others, it becomes unclear as to which method is most appropriate for determining ever-present, omnipresent components of sustainable development.

At this point, sustainable development circumstances must be defined with reference to utility. From this standpoint, sustainable development could be defined as "a condition in which utility levels do not decrease over time." Thus, sustainable society could be thought of as a paradigm for properly managing resources in order to assure that the present average quality of life is accessible to all future generations.

It is important to recognize that sustainable development is not necessarily "a condition in which consumption levels do not decrease over time," which is unrelated to standardized utility. This is due to the fact that utility is directly gained from the environment by every individual and not from commodities produced with natural resources. Therefore, the aim of sustainable development policy is to maintain utility.

At the core of this definition is the pretext that since utility is derived from resources, resource levels should in some way or another be conserved so that they can be passed on to future generations. However, the claim that all conditions should be continuously improved is not justifiable. Instead, sustainable development precepts can be met if aggregated indicators, from natural capital to material and human capital, do not collectively decline. Natural capital includes soil, animals, fish, plants, renewable resources, mineral resources, and other physical assets provided by nature. It can be utilized but not created by humankind. Considering aggregated indicators implies not that decreasing natural capital through use is reprehensible, but that it is possible to compensate for such natural capital reductions with increases in man-made capital. However, if it just so happens that there is no available replacement capital for a given natural capital, then natural capital reductions cannot be compensated for and reductions of said capital should be avoided in order to uphold sustainable development conditions. For example, while wood can be replaced by leather, labor by machines, and metal by plastic, roads cannot replace oak tree forests.

At this point, assume that there are externality costs to excessive resource use and pollution emissions. Here, it is imperative to increase man-made capital levels above natural capital deduction levels accrued through pollution-induced environmental damage. In the same vein, technological innovation is imperative to achieving sustainable development through environmental conservation. In order to achieve said technological development and further its implementation, systems that call for pollution emissions reductions must additionally be linked to corresponding cost reductions. It is precisely through appeals of this sort that such regulations as environmental taxes have come to be highly favored incentivizing factors for regulatory compliance. Furthermore, since determining appropriate prices for externality costs is indeed critical, it follows that sustainable development itself is intricately related to dealings with externality costs.

Table 7.3.1 provides simplified depictions of resource use policies that are crucial for sustainable development. The higher-ranking items are considered, by definition, to be more substitutable, while the lower-ranking items are considered to be less so. Below, we narrow the focus down to linkages between the environment and the

Table 7.3.1 Resource use regulations that promote sustainable development

1	Market failures that arise from environmental and resource pricing must be corrected.
2	Renewable resources must be preserved such that they can continually replenish themselves.
3	Technological progress must be promoted through a system which encourages a transition from using non-renewable to using renewable resources.
4	Renewable resources ought to be used.
5	Economic activity must be restricted to a scope that does not transcend the natural resource limitations.

economy to provide a simplified explanation that does not overemphasize the multi-faceted connections between the economy and environmental and resource factors.

Environmental Kuznets hypothesis

The environmental Kuznets hypothesis describes an inverse U-shaped relationship (the curve that depicts this relationship is known as the *environmental Kuznets curve* that exists between economic development and environmental pollution. That is, although pollution increases at the early stages of economic development, that trend reverses when it surpasses a certain gross domestic product (GDP) per capita, resulting in a reverse trend towards environmental restoration. Through purporting a relationship of improving environmental indicators that are displayed as fixed income levels cross over a transition point, this hypothesis suggests the possibility of sustainable development.

Generally speaking, environmental Kuznets curves are formed as policies clearly recognize such pollutions as SO_2, which itself gives rise to global acid rains and is an underlying factor among many serious health issues to date (Figure 7.3.2 displays the current SO_2 levels throughout the world). In cases concerning pollutants that affect the environment, even if GDP is comparatively low, the effects of curtailing environmental pollution manifest.

Why does this hypothesis hold true? First of all, in the early stage of economic development, it is clear that people ascribe greater value to material affluence than to the environment. People happily accept pollution increases that take place hand in hand with greater consumption. However, as income levels rise, people grow to value the environment more and seek to undo environmental damage even if it means sacrificing a portion of their material affluence. It is at this stage that a clean industrial sector can be structured despite the fact that industry sector activity does traditionally produce pollutants, and tightly regulating these activities could serve to hinder economic growth. If such a clean industrial sector could grow into a main component of economic growth, the host economy could potentially achieve sustainable growth alongside environmental protectionism.

The sustainable development indicated by the Brundtland Committee points to conditions in which conservation alongside environmental preservation leads to more

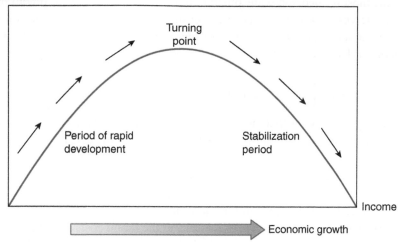

FIGURE 7.3.1 Economic growth and environmental conservation

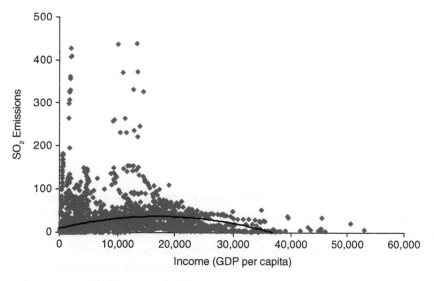

FIGURE 7.3.2 Environmental Kuznets curve—SO_2

sustainable lifestyles among people. Economic development is particularly indispensable to improving the lifestyles of people in developing countries. If the environmental Kuznets hypothesis holds true, then economic development should enable developing countries to reduce pollution emission levels. Under such circumstances, economic cooperation with developed countries in achieving pollution reductions is paramount to effectively diminishing environmental damage. Therefore, based on this hypothesis, it is crucial to determine whether or not developing nations' economic growth could have impacts on reductions among environmental damages. Currently, the most refined and widespread notion of sustainable development measurement that serves as a comprehensive measure for United Nations Sustainable Development Goals is the Inclusive Wealth Index that is published formally within the UN *Inclusive Wealth Report.*

Uncertainty and discount rates

Decision-making for policies that target various environmental issues such as climate change requires due consideration of long-term effects. Specifically, the presence and impacts of future uncertainty heavily weigh in on long-term deliberation. When considering policy oriented towards sustainable development, determining how to think about future uncertainty is of prime importance, as implementing policies that will deal with said uncertainty will lead to long-term impacts.

For example, consider the challenge of measuring the future costs and benefits of an environmental policy. There exists a great deal of uncertainty that is inexorably linked to environmental policies that tie into the availability of future technologies, problems surrounding specific ways to utilize natural resources, and difficulties predicting population migrations.

Chapter 4 highlights the importance of discount rates that affect long-term decision-making. Since it is unknown as to whether or not a discount rate is correct, depending upon what kind of discount is used for long-term issues induces large changes among the effects of uncertainty. When dealing with long-term problems like climate change, if a large discount rate is utilized, any sort of benefit is largely useless. For example, when the discount rate is 5%, the present discounted value (in other words, the benefit) when 1 million yen can be earned after 50 years is just below 90,000 yen, or:

$$1,000,000 / (1 + 0.05)^{50} = ¥87,204$$

In other words, this project would be implemented if 100,000 yen must be paid this year. Now consider a case where present uncertainty prevents the determination of a uniform, optimal discount rate. For example, by estimating two discount rate conditions, 1% and 10%, it is possible to see the probability that they are correct. Note that the present discounted value of each is 600,000 yen and 8,500 yen, respectively. Cases such as this, where there is uncertainty regarding the discount rate itself, greatly influence final decision-making.

The next critical issue is why values lower than the anticipated value of the discount rate should be used in actual policy decision-making when future uncertainty exists. Moreover, in trials with anticipated values, the result is the average numerical value gained. Consider a case where a 1 million yen net benefit is earned after 50 years. Assume that either 0% or 10% is the proper discount rate after 50 years. In this scenario, the anticipated value is 5%, or *(0% + 10%)/2*, and upon calculating this, the value derived from the aforementioned equation is approximately

8,700 yen. However, the net benefit when a discount rate of either 0% or 10% is used becomes 1,000,000 yen and 8,500 yen, respectively. In other words, the anticipated value of the present discounted value, is *(¥91,000,000 + ¥8,500)/2 + ¥504,250*. Or, the anticipated present discount value largely outstrips the present discounted value that utilized the anticipated discount rate. To determine which discount rate it is, we shall now calculate which anticipated present discount value would amount to 504,250 yen. Assuming X stands for the discount rate:

$$1,000,000 / (1 + X)^{50} = 504,250$$

so, $X = (1,000,000/504,250)^{\wedge}1/50 - 1 = 0.014$, and so X is about 1.4%. This is the calculation of the actual approximate discount rate of 1.4% which ought to be utilized in calculations of the actual present discounted value. Here, extreme cases using 0% and 10% were depicted, but in cases where 4% and 6% are utilized, the numbers drop below just beneath 5%. The main point is recognizing the big problem of choosing a discount rate when there is uncertainty with the discount rate. Furthermore, one should understand the large impact on present benefits that the chosen discount rate has.

SUMMARY

Sustainable development can be defined as a condition in which efficiency standards do not decrease over time. Furthermore, conditions that do not lead to decreases in aggregated natural, human, and material capital satisfy sustainable development conditions.

REVIEW PROBLEMS

1. Define sustainable development.
2. Explain what comprises the environmental Kuznets curve.
3. Explain what kind of discount rate should be used in the presence of discount rate uncertainty.

LEARNING POINT: ENVIRONMENTAL KUZNETS CURVE

Although CO_2, as a catalyst of climate change, will impose serious impacts on future generations, the inability to concretely connect it to specific future environmental issues remains an issue. When emissions levels across the planet become problematic, pollution alongside income increases becomes monotonic, as depicted global CO_2 conditions in Figure 7.3.3. This is said to not conform to the environmental Kuznets hypothesis. Furthermore, even if it does conform, since GDP is prioritized over pollution emission restrictions, if GDP levels are not relatively high, then it is thought that this means that pollution emissions restrictions are not yielded.

There are countries where a long period of time is needed to reach the turning point on an environmental Kuznets curve, since incomes are extremely low, poverty issues are prioritized over environmental issues, and there is a high chance for pollution issues to be neglected as economic development proceeds. For these countries, through economic cooperation, policies are thought to be important for resolving poverty issues and reducing burdens on the environment. And since economic growth in some African and other countries will actually be negative, it is essential to understand that if things continue as they are, there will be countries that are not even able to reach the transition point on the Kuznets curve.

Now, the Kuznets hypothesis is often used in judgments for rationalizing environmental degradation through economic growth. If the hypothesis does not hold true, then there are two choices available: either abandon economic development for environmental protectionism, or continue to bolster the economy regardless of concomitant environmental destruction. In reality, it is nearly impossible for not only developing countries, but also developed countries to give up on their pursuits of economic growth. That being said, however, without environmental preservation efforts, sustainable economic development is also nearly impossible. The environment could be said to be the foundation of society that props up economic activity. By not preserving it, the environment would degrade to the critical state represented

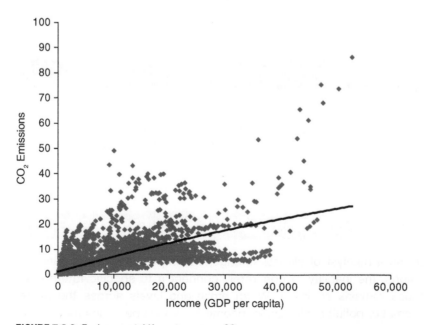

FIGURE 7.3.3 Environmental Kuznets curve—CO_2

by *C* in Figure 7.3.4, leading to irreversible damages and the collapse of society. This trajectory towards *C* is the road to the collapse of society. Furthermore, in scenarios where there were more pollution than a fixed amount, income levels would not necessarily approach zero, but further economic development would become impossible. In other words, the developmental limit would be reached.

Easter Island is an example of this type of failure. At one time, Easter Island was completely covered by lush, green forestry, and lumber resources were largely consumed by a prosperous civilization. Specifically, this forest lumber was harvested in order to construct housing, ships, and Moai statues. Unplanned tree felling led to the complete destruction of the forests, which subsequently contributed to food shortages, bitter civil wars among starving peoples, and the fall of the Easter Island civilizations. This is known as a case were environmental destruction undermined the environment's natural ability to replenish itself (e.g. its resources).

On the other hand, Japan's Edo Period is a more promising example. Just like Easter Island, the demand for timber resources skyrocketed alongside a rush to construct castles and housing throughout all parts of Edo Period Japan. Although the Tokugawa Shogunate faced the dangers of forest destruction, it responded swiftly, implementing a wide range of forest harvest supply restrictions. This led to it being the first society of its time to successfully avoid the danger of civilization collapse by measuring the increasing supplies of forest resources and responding appropriately. This is said to be a case where before the environment's ability to naturally replenish itself was negatively impacted, environmental policies were appropriated in order to induce transitions in the activities of the manufacturing sector so that environmental destruction did not come to be. In other words, to make economic growth and environmental preservation possible simultaneously, not only simple income amalgamations, but also the environment's ability to rejuvenate itself as well as pertinent policies and technologies are of prime importance.

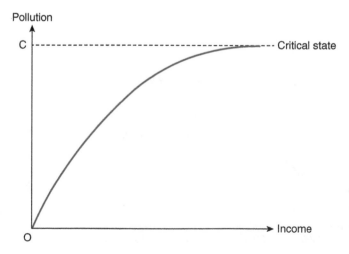

FIGURE 7.3.4 Societal collapse

Section 7.4: Energy economics

World energy

While energy is indispensable to our lives, its use is often linked to greenhouse gas emissions and air pollution that are detrimental to the ecosystem and exacerbate climate change. In developed countries, businesses and governments put forth environmental countermeasures, and by restricting the use of energy that brings about large greenhouse gas and air pollution emissions, they are working towards efficient energy use. However, as demand for developing countries is expected to escalate hand in hand with economic development and population increases, worldwide energy consumption levels will continue to increase alongside economic development.

If climate change policy continues in its present form (if current policies are promoted), it is estimated that increases in global energy demand will increase in a trend similar to the thick line depicted in Figure 7.4.1. World energy demand has continually increased by 2.5% per year, from 3.8 billion tons of oil equivalent (t.o.e.) in 1965 to 11.2 billion t.o.e. by 2009, three times the former levels. Even so, through the promotion of climate change policies, it is possible that overall energy demand levels can be reduced. In order to suppress future atmospheric temperature rises to within 2 degrees (based on global temperature averages of the pre-Industrial Revolution era), a range where the effects of climate change would be small, future energy use must be curtailed by up to 20%. In the graph, the energy transitions needed to achieve this goal are represented by the dotted line.

Moreover, energy demand varies regionally. At the heart of the Asia/Oceania region, marked by considerable economic development across the board, the rate of increases in energy demand among developing countries is high (from 1965 to 2009, the demand

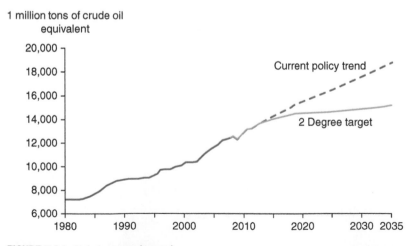

FIGURE 7.4.1 Global energy demand

multiplied by 8.5 times, reaching 4.1 billion t.o.e.). Specifically, the majority of future energy consumption increases are expected to take place among non-OECD developing countries, where some 2 billion people reside without access to electricity or city gas. Furthermore, since energy conservation has progressed among developed countries, their energy demand increase rate will have a declining trend.

Like any other resource, the price for energy is determined by the effects of supply and demand (refer to Chapter 1, Section 3). If policy standards were maintained in their current state, it is possible that prices could rise alongside increases in global energy demand, or fall with demand plunges. We know from the example of oil price movements depicted in Figure 7.4.2 that there have been significant fluctuations. In the scenario where energy demand continues to increase by current trends and oil prices subsequently increase, it is estimated that oil prices will rise by up to 50% of current levels. However, if energy demand is successfully suppressed to levels that could allow the "2 degree target" to be realized, then prices in the ballpark range of current levels could potentially be preserved. Moreover, large fluctuations in oil prices are also influenced by the financial flows and political processes of the gold market.

At this point, take a look at the demand for each global energy source, both present and future. Oil is at the center of energy consumption up until now, making up a larger than 30% share of total energy consumption. This is due to the large increase in oil consumption as a transportation fuel, and the fact that the fulfillment of this function has not been replaced by other energy sources. Next, coal and natural gas also occupy large portions of global consumption. Demand for coal as a cheap electricity generation fuel in China and other developing Asian countries has led to its increasing consumption rates. And, as greenhouse gas emissions from natural gas use are less than those from oil use, demand in developing countries aiming to cope with climate

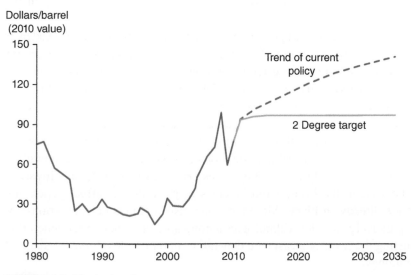

FIGURE 7.4.2 Oil price trends

change is increasing. Finally, nuclear power and renewable energy occupy a small share of total consumption, yet have relatively large proliferation rates.

The IEA World Energy Outlook depicts incremental increases in demand levels from the year 2009 through the year 2035. The aforementioned scenario where current policy trends occur, as well as policies that achieve the "2 degree targets," are both represented as differing increment levels through 2035. Usage amounts of renewable energies including wind power, solar light, solar heat, snow and heat use, temperature difference use, and others, are anticipated to increase alongside natural gas through the year 2035. This indicates that, while the renewable energies currently account for a small share of global energy demand, there is potential for them to play important roles in the future.

Renewable energy in Japan

In Japan and other places, one of the major reasons that renewable energies have not been implemented more than non-renewable energies is because their production costs are so high. Because production costs are high, they lose price competition, which hinders business' incentives to invest in such facilities and research and development. However, as renewable energies are sources with little negative impact on the environment, reducing production and distribution costs through high diffusion rates, alongside promoting technological development, is ideal. In order to promote climate change countermeasures, secure energy resources, deal with pollution, and reduce the costs of renewable energy proliferation, Japan and many other countries have adopted *feed-in tariff* systems that legally stipulate energy purchase prices. Within these systems, the government reflects upon the present electricity generation costs and decides the transaction prices for electricity generated by solar, wind, biomass, and other renewable sources. Then, local energy utilities are obliged to purchase electricity generated from renewable energy facilities operating in the same locality.

Through making the sale price of power generated by renewable energies fixed over the long term by installing generation facilities, it becomes easier to reach goals for returns on facility investment costs, effectively promoting investment and financing. Therefore, feed-in tariff systems are structures that can make cheap supply of renewable energy possible. It is also considered to promote energy production efficiency improvements after generation plants are installed, leading to production cost reductions. The system accomplishes this by setting relatively high energy sales prices for businesses that adopt target technologies for the first time and by reducing the financial assistance (i.e. legally-backed high sales prices) as time goes on. However, revisions of the system that bring about changes in price do not affect generation facilities that are already in place. Moreover, this also includes considerations for diffusing comparatively superior technologies among solar light and other generation forms by adjusting the level of assistance depending upon the development stage of the given technology.

While feed-in tariff systems promote the expansion of renewable energy through fixing energy prices, it is also possible to promote market proliferation by standardizing energy use levels. Systems that oblige energy users to consume renewable energy as a pre-determined proportion of their total energy consumption are known as *renewables portfolio standard* (RPS) systems. A company can fulfill such obligations by producing its own renewable energy-based power or through purchasing a suitable amount of renewable energy-based power from another producer who generates renewable energy that exceed baseline RPS requirements. Japan had an RPS system in place as of April, 2003, and it switched to a FIT system in July, 2012.

Deregulations in the electricity and gas industries

Electricity industry

Since there was no competition among the ten regional monopoly electricity utilities in Japan, in the latter half of the 1990s, competition fundamentals began to be implemented in stages, beginning with deregulation. First, in 1995, with regards to the procurement of electricity resources of such electric utilities as Tokyo Electric Power Company and Kansai Electric Power Company, a competitive bidding scheme was implemented with *independent power producers (IPPs)* who could supply wholesale electricity. Competitive bidding is a process through which those who desire a sale or to undertake a contract submit a form with a bid price written on it, and from the contents and requested price, acceptance or refusal of the contract is decided. Through this process, contracts could be executed with businesses that suggest the most favorable conditions. Vending, gas, cement, and other new enterprises participated as new entrants, and purchasing from IPPs during electricity supply shortages due to the impacts of the Great East Japan Earthquake on March 11th, 2011, reinforced the system.

Beyond this, in line with operational changes within the *Electric Utility Industry Law*, a fee system based upon the *yardstick assessment method* was put into effect. For supplying household demand, which was not targeted by electricity deregulation, the ten regional monopoly electricity utilities were taken into account, and conditions that would promote indirect competition among the utilities that had no direct competitive relations were estimated. This thereby led to the establishment of a comparative evaluation system that aimed to measure the cost reductions, efficiency, and other targets that existed in common across all of the electric utilities. In other words, this system was anticipated to induce competition in price reduction attempts as the lowest operating costs among the group of utilities came to be referred to as the industry standard.

Later, in the year 2000, retail sales in demand areas of over 2,000 kW receiving electricity via greater than 20,000 V power lines were deregulated. For these demand centers, *power producer and suppliers (PPS)* were at once able to join the market as electricity vendors. Furthermore, ordinary electricity businesses and the wholesale electricity businesses were both permitted to bid for thermal power generation with the new slate of participants. Through regulation changes such as this, and subsequently, the revisions made to the *Electric Utility Industry Law* in the year 2003, further liberalizations came into effect in consecutive, gradual stages.

Gas industry

The city gas industry in Japan is comprised of general gas companies (209 companies as of March, 2012) and community gas utilities (1,475 companies as of March, 2012). The major special feature of the Japanese gas industry is that each region is not divided and covered by a few companies, as is the case with the electricity industry. Instead, a variety of both large and small enterprises operate independently. Because of that, relative to the Japanese territory, the number of companies is large. General gas utilities include both city gas companies and companies that supply gas through ducts depending on general demand. Furthermore, community gas utilities are required by law to make use of simple, easy-to-provide products and equipment such as LPG stored in cylinders.

Gradual deregulation of the gas industry in Japan began in the latter half of the 1990s. First, in March, 1995, the regulations and financial barriers to supplying large gas consuming demand centers, specifically those related to large-scale business facilities that purchase over 2 million m^3 of gas per contract year, were eased. This legal revision was the first of its kind in a 25-year time span since 1970. Unfortunately, the revision failed to become a revolutionary change for large-scale consumers, as the scope of the framework was limited to consumers of over 2 million m^3 of gas. This point was addressed, however, by further system revisions adopted in November, 1999 which expanded the definition of "large-scale consumers" to city hotels and other organizations that consumed at least 1 million m^3 per year. And, aiming to further promote competition after that, the reforms implemented in April, 2004 expanded the scope to offices consuming 500,000 m^3 per year. Finally, in April, 2007, supply to merchant facilities using over 100,000 m^3 per year was liberalized. Through this, the share of new participants has grown over the years.

SUMMARY

Now and into the future, global energy consumption levels will continually increase alongside economic development. To date, crude oil has been the main global fuel source at the core of primary energy consumption. Recently renewable energies that make up small shares of total consumption are gaining popularity and being more widely utilized. Additionally, in Japan as well as many other countries, feed-in tariffs that set energy transaction prices by law are gaining in prominence. Furthermore, future energy systems are expected to evolve and transform as new power producers and suppliers enter increasingly deregulated electricity markets.

REVIEW PROBLEMS

1. Describe future global energy resources.
2. Explain how Japan has successfully promoted renewable energy proliferation.
3. Describe energy source options and their various benefits and drawbacks.

LEARNING POINT: NUCLEAR POWER

At 2:46 p.m. on March 11th, 2011, the magnitude 9.0 Great East Japan Earthquake occurred in Sanriku waters off the Pacific coast of Japan. Reaching a 6.0 maximum seismic intensity, the massive quake wrought large-scale damage throughout Japan, from Tohoku to Kanto regions. The quake and its subsequent tsunami brought about the infamous Fukushima Daiichi Nuclear Power Station accident, leading Japan to its current issue with restarting its shutdown nuclear power plants. Since then, the safety of nuclear reactors, and how limits on reactor lifespans impact the number of nuclear reactors have been debated. Here, we will introduce lifespans and systems that should lead to a nuclear reactor's decommissioning.

Modern nuclear reactors are constructed with an anticipated lifespan of 30 to 40 years before decommissioning, so when commonly used infrastructure is installed altogether at once, it can be used for a long time. Naturally, 60-year lifespan reactors are used for longer than 40-year models. There is a great discrepancy between curtailment trends between the U.S. and Japan. As many as four reactors have been newly constructed in the U.S. since the 1990s after the 1979 Three Mile Island incident that took place in Pennsylvania, U.S.A., and the 1986 Chernobyl incident that occurred in the Soviet Union.

Thus, while the U.S. is currently the country with the world's most nuclear power plants (104 in total), if it were not to construct any new reactors from now on, it would only have two by the year 2030. As it is believed that the next generation nuclear reactors will become commercially viable in the latter half of the 2020s, implementing replacements for current nuclear reactors that will go out of use has become a pertinent topic of discussion. Furthermore, Japan, similar to the U.S., had been constructing new reactors every year, and out of the 40-year lifespan facilities, 35 will remain by the year 2020, while only 19 will still be in use by the year 2030.

Therefore, when it comes to creating new systems, energy choices based on nuclear power plants and other costs borne by producers, as well as those based on consumer decision-making, are in high demand. The costs that businesses ought to bear are not only those which impose burdens on the environment as have been outlined in this unit thus far, and also include insurance fees for nuclear power accident compensation arrangements, backend costs, system and regulation costs, and other expenses. Backend costs are those related to the reprocessing of spent nuclear fuel, decommissioning factories, processing nuclear waste, and so on.

System fees include the culmination of power generation, transformation, transmission, and distribution costs necessary to supply electricity to power incoming units for household demand. By including all of these into the cost calculations of nuclear power, it is possible to accurately determine actual electricity costs.

Feed-in tariffs and other systems have been used to promote the use of costly energy sources, and beyond that, the energy industry has undergone regulatory easing. The ability for consumers to choose appropriate energies is of prime importance, and the ideal percentage of total energy consumption that specific energy sources make up is not the true issue at hand. Consumers who hold anti-nuclear, pro-renewable energy sentiments can pour their support into renewable energy providers by purchasing from them, whilst those who believe that nuclear power is a noteworthy component of climate change policy can purchase energy from nuclear power providers, and those who are most concerned with energy prices can buy from producers who charge the least amount of money. Thus, in the end, the best energy mix will arise from producers who provide energy in line with consumer support.

Bibliography

Basel Convention on the Control of Transboundary Movements of Hazardous Wastes and the Disposal, Basel, 22 March 1989. *United Nations Treaty Series*, vol. 1673, p.57. https://treaties.un.org/pages/ViewDetails.aspx?src=TREATY&mtdsg_no=XXVII-3&chapter=27&clang=_en.

Clean Development Mechanisms Executive Board (CDM EB). http://cdm.unfccc.int/EB/index.html.

Coase, R., *The Problem of Social Cost, Journal of Law and Economics*, Vol. 3(1): pp. 1–44, 1960.

Contribution of Working Group 1 to the Fourth Assessment Report of the Intergovernmental Panel on Climate Change, IPCC, Geneva, Switzerland, 2007.

Convention on International Trade in Endangered Species of Wild Fauna and Flora (Washington Convention). Washington. 3 March 1973. *United Nations Treaty Series*, vol. 993. https://treaties.un.org/Pages/showDetails.aspx?objid=0800000280105383.

Council for PET Bottle Recycling, accessed March 26, 2016, http://www.petbottle-rec.gr.jp/english/.

European Commission. "The European Union Emissions Trading System (EU ETS)." http://ec.europa.eu/clima/policies/ets/index_en.htm.

European Union Emissions Trading System, accessed April 4, 2016, http://ec.europa.eu/clima/policies/ets/pre2013/documentation_en.htm.

Hardin, Garrett. "The Tragedy of the Commons." *Science* 162 (1968): 1243–1248.

International Energy Agency. *CO2 Emissions from Fuel Combustion Highlights 2015*. http://www.iea.org/statistics/.

IEA, *Emissions from Fuel Combustion,* International Energy Agency, Paris, 2011.

IEA, *World Energy Outlook 2012*, International Energy Agency, Paris, 2012.

International Union for Conservation of Nature and Natural Resources. *World Conservation Strategy*. 1980. https://portals.iucn.org/library/efiles/html/wcs-004/cover.html.

Itsubo Norihiro, Inaba Astushi (eds). *Life Cycle Environmental Impact Methods: LIME-LCA, Environmental Accounting, Database and Assessment Methods for Environmental Efficiency*, Maruzen, 2005.

Japan Federation of Economic Organizations, *Japan's Response to Global Warming*, September 19, 2001.

Japan Standards Association, Secretariat of the Japanese National Committee for ISO TC 207, http://www.jsa.or.jp/default_english/default_english.html.

Kyoto Protocol to the United Nations Framework Convention on Climate Change. Dec. 10, 1997. U.N. Doc. FCCC/CP/1997/7/Add.1.37 I.L.M.22 (1998).

Kuriyama, Koichi. "Public Works and Environmental Assessments: The Role of Cost-Benefit Analysis in Environmental Assessments", *Japanese Annual Report on Environmental Economics & Policy*, pp. 55–67, 2003.

Law for the Promotion of Sorted Collection and Recycling of Containers and Packaging, 2006.

Meadows, D.H., & Club of Rome, *The Limits to Growth: A Report for the Club of Rome's project on the predicament of mankind*, New York: Universe Books, 1972.

Meadows, Donella H., Meadows, Dennis L., Randers, Jorgen. *Beyond the Limits*. Chelsea Green Publishing, 1992.

Ministry of Agriculture, Forestry, and Fisheries, Government of Japan. *2011 White Paper on Fisheries*, 2011.

Ministry of Economy, Trade, and Industry, Government of Japan, accessed March 31, 2016, http://www.meti.go.jp/english/.

Ministry of Economy, Trade, and Industry, Industial Structure Council, Environment Section, Government of Japan. http://www.meti.go.jp/committee/kenkyukai/energy_environment.html. [Japanese]

Ministry of Economy, Trade, and Industry, Government of Japan, *Top Runner Program,* 1998.

Ministry of Environment, Environmental Economics Department, Environmental Policy Bureau, Government of Japan, *Survey of Environmentally Friendly Business Operations*, 2012.

Ministry of Environment, Government of Japan, *Annual Report on Environmental Statistic*s, 2012.

Ministry of Environment, Government of Japan. *Annual Report on the Environment, the Sound Material-Cycle Society and Biodiversity in Japan 2011*. http://www.env.go.jp/en/wpaper/.

Ministry of Environment, Global Warming Tax Committee, Government of Japan. http://www.env.go.jp/council/16pol-ear/yoshi16-01.html. [Japanese]

Ministry of Internal Affairs and Communications, Statistics Bureau, Government of Japan. *Japan Series of Long-Term Statistics*, http://www.stat.go.jp/english/data/index.htm.

Organization for Economic Cooperation and Development (OECD) Development Centre Studies, *The World Economy, Volume 2: Historical Statistics*, 2006.

Slovic, P. "Perception of Risk," *Science,* Vol. 236, pp. 280–285, 1987.

Statistics Bureau of Japan. 日本の長期統計系列. 2014. http://www.stat.go.jp/data/chouki/. [Japanese,no longer in publication]

The Economics of Ecosystems and Biodiversity (TEEB), *The Economics of Ecosystems and Biodiversity: The Ecological and Economic Foundations*, TEEB D0, 2009.

United Nations Conference on Environment and Development. *Agenda 21*. 14 June 1992. https://sustainabledevelopment.un.org/outcomedocuments/agenda21.

United Nations Environment Programme. *Rio Declaration on Environment and Development*. 14 June 1992. http://www.unep.org/documents.multilingual/default.asp?documentid=78&articleid=1163.

United Nations Framework Convention on Climate Change, Clean Development Mechanisms Executive Board, Kyoto Mechanisms Information Platform, 2012.

United Nations Millennium Ecosystem Assessment, *Ecosystems and Human Well-Being*, 2007.

United Nations University International Human Dimensions Programme on Global Environment Change, United Nations Environmental Program, *Inclusive Wealth Report 2014: Measuring Progress Towards Sustainability*, Cambridge University Press, 2014.

United Nations World Commission on Environment and Development (Brundtland Commission). *Our Common Future*. 1983. https://sustainabledevelopment.un.org/milestones/wced.

Viscusi, W.K., Hakes, J. and Carlin, A., "Measures of Mortality Risks," *Journal of Risk and Uncertainty*, Vol. 14, No. 3, pp. 213–233, 1997.

Watanabe, M., "Trade Deregulation and Environmental Problems under the WTO", *RIM* No. 35, 1996.

World Wildlife Fund, *Living Planet Report*, 2012.

Index